The **Chaotic Pendulum**

The Chaotic Pendulum

Moshe Gitterman

Bar-Ilan University, Israel

 **World Scientific**

NEW JERSEY · LONDON · SINGAPORE · BEIJING · SHANGHAI · HONG KONG · TAIPEI · CHENNAI

Published by

World Scientific Publishing Co. Pte. Ltd.

5 Toh Tuck Link, Singapore 596224

USA office: 27 Warren Street, Suite 401-402, Hackensack, NJ 07601

UK office: 57 Shelton Street, Covent Garden, London WC2H 9HE

British Library Cataloguing-in-Publication Data
A catalogue record for this book is available from the British Library.

ISBN-13 978-981-4322-00-3
ISBN-10 981-4322-00-8

Printed in Singapore.

Preface

The second part of the title of this book is familiar to everybody who swings
to and fro in childhood, or has changed his/her mind back and forth from joy
to grief in mature age, whereas the first part of the title needs explanation.
We use the word "chaotic" as a synonym for "unpredictable." Everything
was clear until the third quarter of the last century: all phenomena in
Nature were either deterministic or chaotic (random). The solution of a
second-order differential equation contains two arbitrary constants which
can be found from the initial conditions. By defining the coordinate and
velocity of the particle at time $t = 0$, one can calculate these variables
deterministically at each later time $t > 0$. On the other hand, if a system,
say, a Brownian particle, is subject to a random number of collisions with
small particles, its motion will be non-predictable (random). However,
everything changed in the 1970s, as can be seen from the title of a 1986
conference on "chaos": "Stochastic Behavior Occurring in a Deterministic
System." Although the idea of "deterministic chaos" appeared earlier, the
intensive study of this phenomenon only started about forty years ago. The
answer to the question in the title of the article [1], "Order and chaos: are
they contradictory or complimentary?", is now obvious.

The concept of "chaos" is usually associated with systems having a large
number of degrees of freedom. The approach of statistical mechanics allows
one to calculate the average characteristics of the system, leaving the behavior of individual particles as "random". It turns out, however, that chaos
may appear in a differential equation with only three variables, provided
that the system is nonlinear. It is crucial to distinguish between linear and
nonlinear differential equations. An important property of chaotic nonlinear equations is the exponential increase in time of their solutions when
one makes even the smallest change in the initial conditions. "Deterministic

chaos" appears without any random force in the equations. Such a situation is very common since an infinite number of digits is required to specify
the initial conditions precisely, an accuracy that is obviously unattainable
in a real experiment. The exponential dependence on initial conditions is
popularly known as the "butterfly effect," which means that an infinitesi-
mal change in initial conditions produces very different results in the long
run (the flapping of a butterfly's wing in Texas may create large changes
in the atmosphere leading to a tornado in the Pacific Ocean). A very small
change in initial conditions can transform a deterministic system into a
chaotic system. Therefore, considering "chaotic motion" means considering the general properties of nonlinear differential equations (Chapter 2),
as well the effect of a random force (Chapter 3). The chaotic behavior of
a spring, double and spherical pendula are the subject of Chapter 4. A
general introduction to the subject can be found in my previous book [2].

Two main features characterize the study of chaos. First, nonlinear
dynamics and chaos are an area of intensive mathematical investigation.
Second, due to the lack of analytical solutions, this field is usually studied
by numerical methods. This book does not contain rigorous mathematical
statements or details of numerical methods related to chaos. My aim was to
make the presentation as simple as possible, so that a scientist or a student
having only a general knowledge of mathematical physics, could easily find
in this small volume all the required information for their theoretical or
laboratory work.

Contents

List of Equations

$$\frac{d^2\phi}{dt^2} + \alpha\frac{d\phi}{dt} + [A\sin(\Omega t) + \xi(t)]\sin\phi = 0$$

$$\frac{d^2\phi}{dt^2} + \alpha\frac{d\phi}{dt} + [1 + A\sin(\Omega t)]\sin\phi = \xi(t)$$

$$\frac{d^2\phi}{dt^2} + \frac{d\phi}{dt} + [\omega_0^2 + \xi(t)]\sin\phi = a + f\sin(\Omega t) + \eta(t)$$

Overdamped pendulum

$$\frac{d\phi}{dt} = a_0 - b_0\sin\phi + \xi(t)$$

$$\frac{d\phi}{dt} + b_0\sin\phi = \xi(t)$$

$$\frac{d\phi}{dt} = a_0 - [b_0 + \eta(t)]\sin\varphi$$

$$\frac{d\phi}{dt} = [a_0 + \xi(T)] - [b_0 + \eta(t)]\sin\varphi$$

$$\frac{d\phi}{dt} = [a_0 + \xi(T)] - [b_0 + \eta(t)]\sin\varphi$$

$$\frac{d\phi}{dt} = [a_0 + \xi(T)] - [b_0 + \eta(t)]\sin\varphi$$

Chapter 1

Pendulum Equations

1.1 Mathematical pendulum

The pendulum is a massless rod of length l with a point mass (bob) m at its end (Fig. 1.1). When the bob performs an angular deflection ϕ from the equilibrium downward position, the force of gravity mg provides a restoring torque $-mgl\sin\phi$. The rotational form of Newton's second law of motion states that this torque is equal to the product of the moment of inertia ml^2 times the angular acceleration $d^2\phi/dt^2$,

$$\frac{d^2\phi}{dt^2} + \frac{g}{l}\sin\phi = 0 \tag{1.1}$$

Fig. 1.1   Mathematical pendulum.

For small angles, $\sin\phi \approx \phi$, Eq. (1.1) reduces to the equation of a harmonic oscillator. The influence of noise on an oscillator has been considered earlier [2]. The main difference between the oscillator and the pendulum is that the former has a fixed frequency $\sqrt{g/l}$, whereas the pendulum period decreases with increasing amplitude. Multiplying Eq. (1.1) by $d\phi/dt$ and integrating, one obtains the general expression for the energy of the pendulum,

$$E = \frac{l^2}{2}\left(\frac{d\phi}{dt}\right)^2 + gl\left(1 - \cos\phi\right) \qquad (1.2)$$

where the constants were chosen to make the potential energy vanishes at the downward vertical position of the pendulum. Systems with constant energy are called conservative systems. In the $(\phi, d\phi/dt)$ plane, the trajectories are contours of constant energy.

Depending on the magnitude of the energy E, there are three different types of phase trajectories in the $(\phi, d\phi/dt)$ plane (Fig. 1.2):

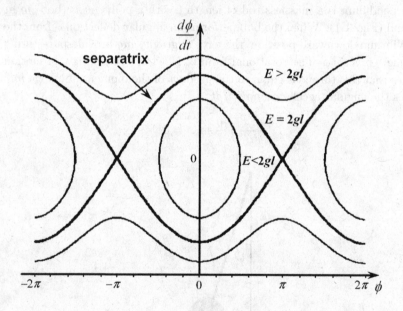

Fig. 1.2   Phase plane of a mathematical pendulum.

1. $E < 2gl$. The energy is less than the critical value $2gl$, which is the energy required for the bob to reach the upper position. Under these conditions, the angular velocity $d\phi/dt$ vanishes for some angles $\pm\phi_1$, i.e., the pendulum is trapped in one of the minima of the cosine potential well, performing simple oscillations ("librations") around the position of the min-

imum. This fixed point $\phi = d\phi/dt = 0$ is called an "elliptic" fixed point, since nearby trajectories have the form of ellipses.

2. $E > 2gl$. For this case, there are no restrictions on the angle ϕ, and the pendulum swings through the vertical position $\phi = \pi$ and makes complete rotations. The second fixed point $(\phi = \pi, d\phi/dt = 0)$, which corresponds to the pendulum pointing upwards, is a "hyperbolic" fixed point since nearby trajectories take the form of hyperbolas. It is sufficient to study these two equilibrium points $\phi = 0$ and $\phi = \pi$ since all other fixed points $\phi = n\pi$ with $|n| > 1$ can be deduced by periodicity.

The orbits shown in Fig. 1.2, which are periodic in time, are called "limit cycles." For small energies close to the origin, the orbits are simple harmonics with a single peak in the power spectrum. For larger energies, the power spectrum also contains different harmonics of the main frequency. The special orbits are either "homoclinic," which leave the hyperbolic fixed point and then return to it, or "heteroclinic," which connect two hyperbolic fixed points. In the hyperbolic fixed points, the stable manifold merges with the unstable manifold, thus forming the heteroclinic orbit. This orbit shows the exchange stability between oscillatory and rotating motion ("heteroclinic bifurcation").

The important property of a Hamiltonian system (with no damping) is "conservation of volume in phase space." This means that initial conditions, distributed in some volume, will remain in the same volume with time. Hamiltonian systems cannot have attractors which attract different trajectories since, otherwise, the initial conditions distributed inside some finite volume will reduce to the zero volume occupied by an attractor.

3. $E = 2gl$. For this special case, the pendulum reaches the vertical position $\phi = \pi$ with zero kinetic energy, and it will remain in this unstable point until the slightest perturbation sends it into one of the two trajectories intersecting at this point. The border trajectory, which is located between rotations and librations, is called the separatrix, since it separates different types of motion (oscillations and rotations). The equation of the separatrix can be easily obtained from (1.2),

$$\frac{d\phi}{dt} = 2\sqrt{\frac{g}{l}} \cos\frac{\phi}{2} \tag{1.3}$$

The time t needed to reach the angle ϕ is given by

$$t = \sqrt{\frac{l}{g}} \ln\left[\tan\left(\frac{\phi}{4} + \frac{\pi}{4}\right)\right] \tag{1.4}$$

Trajectories close to the separatrix are very unstable and any small perturbations will result in running or locked trajectories. These trajectories possess interesting properties [3]. Two trajectories intimately close to the separatrix, with energies $0.9999(2gl)$ and $1.0001(2gl)$, describe the locked and running trajectories, respectively. In spite of having almost the same energy, their periods differ by a factor of two (!) so that the period of oscillation is exactly twice the period of rotation. Physical arguments support this result [3].

It is convenient to perform a canonical transformation from the variables ϕ and $d\phi/dt$ to the action-angle variables J and Θ [4]. For librations, the action J is defined as

$$J = \frac{1}{2\pi} \oint \left(\frac{d\phi}{dt}\right) d\phi = \frac{\sqrt{2}}{l} \int_{-\phi_1}^{\phi_1} d\phi \sqrt{E - gl + gl\cos\phi}$$

$$= \frac{8\sqrt{gl}}{\pi} [E(\kappa) - \kappa^2 K(\kappa)] \tag{1.5}$$

where $K(\kappa)$ and $E(\kappa)$ are the complete elliptic integrals of the first and second kind with modulus $\kappa = \sqrt{E/2gl}$. The angle Θ is defined by the equation

$$\frac{d\Theta}{dt} = \frac{\partial E}{\partial J} = \frac{\pi\sqrt{gl}}{2K(\kappa)}, \tag{1.6}$$

yielding

$$\Theta(t) = \frac{\pi\sqrt{gl}}{2K(\kappa)}t + \Theta(0) \tag{1.7}$$

One can easily find [4] the inverse transformation from (J, Θ) to $(\phi, d\phi/dt)$,

$$\phi = 2\sin^{-1}[\kappa sn(2K(\kappa)\Theta/\pi, \kappa)]; \qquad \frac{d\phi}{dt} = \pm 2\kappa\sqrt{gl}cn(2K(\kappa)\Theta/\pi, \kappa)$$
$$\tag{1.8}$$

where sn and cn are Jacobi elliptic functions.

For the case of rotations, there is no turning point, but one can define the action J for the running trajectory as

$$J = \frac{1}{2\pi} \int_{-\pi}^{\pi} d\phi \sqrt{2(E - gl) + gl\cos\phi} = \frac{4\sqrt{gl}}{\pi\kappa} E(\kappa_1) \tag{1.9}$$

where the modulus $\kappa_1 = 2gl/E$. The angle Θ, is obtained as in the previous case,

$$\Theta(t) = \frac{\pi\sqrt{gl}}{\kappa_1 K(\kappa_1)}t + \Theta(0) \tag{1.10}$$

A canonical transformation to the original variables yields

$$\phi = 2am\left(\frac{K(\kappa_1)\Theta}{\pi}, \kappa\right); \qquad \frac{d\phi}{dt} = \pm 2\frac{\sqrt{gl}}{\kappa_1}dn\left(\frac{K(\kappa_1)\Theta}{\pi}, \kappa\right) \tag{1.11}$$

where am is the Jacobi elliptic amplitude function, and dn is another Jacobi elliptic function.

1.2 Period of oscillations

One can easily write the solution of Eq. (1.1) for ϕ and $d\phi/dt$ in terms of elliptic integrals. Since the energy is conserved, Eq. (1.2) becomes

$$\left(\frac{d\phi}{dt}\right)^2 = -\frac{2g}{l}(1 - \cos\phi) + Const. = -\frac{4g}{l}\sin^2\left(\frac{\phi}{2}\right) + Const. \tag{1.12}$$

Denoting the value of $(d\phi/dt)^2$ in the downward position by A, and $\sin(\phi/2)$ by y, one can rewrite (1.12),

$$\left(\frac{dy}{dt}\right)^2 = \frac{1}{4}(1 - y^2)\left(A - \frac{4g}{l}y^2\right) \tag{1.13}$$

Consider separately, the locked and running trajectories are those for which the bob performs oscillations and rotations, respectively, around the downward position. In the former case, $d\phi/dt$ vanishes at some $y < 1$, i.e., $Al/4g < 1$. Introducing the positive constant k^2 by $A = 4gk^2/l$, one can rewrite Eq. (1.13),

$$\left(\frac{dy}{dt}\right)^2 = \frac{g}{l}(1 - y^2)(k^2 - y^2). \tag{1.14}$$

The solution of this equation is [5]

$$y = ksn\left[\sqrt{\frac{g}{l}}(t - t_0), k\right] \tag{1.15}$$

where sn is the periodic Jacobi elliptic function. The two constants t_0 and k are determined from the initial conditions.

For the running solutions $Al/4g > 1$, for which $k < 1$, the differential equation (1.14) takes the following form,

$$\left(\frac{dy}{dt}\right)^2 = \frac{gl}{k^2}\left(1 - y^2\right)\left(1 - y^2 k^2\right) \qquad (1.16)$$

The solution of this equation is

$$y = k\,sn\left[\sqrt{g/l}\frac{t - t_0}{k}, k\right] \qquad (1.17)$$

Finally, for $Al = 4g$, the bob just reaches the upward position. In this case, Eq. (1.16) takes the simple form,

$$\left(\frac{dy}{dt}\right)^2 = \frac{g}{l}\left(1 - y^2\right)^2 \qquad (1.18)$$

whose solution is

$$y = \tanh\left[\sqrt{g/l}\left(t - t_0\right)\right]. \qquad (1.19)$$

One can use dimensional arguments [6] to find the period of oscillation of the pendulum. Equation (1.1) contains only one parameter, $\sqrt{g/l}$, having dimensions of inverse time. Therefore, the product of T and $\sqrt{g/l}$ is dimensionless,

$$T\sqrt{\frac{g}{l}} = f\left(\phi\right) \qquad (1.20)$$

For small angles, $\sin\phi \approx \phi$, and the pendulum equation (1.1) reduces to the simple equation of harmonic oscillator with the well-known solution $T_0 = 2\pi\sqrt{l/g}$, corresponding to $f\left(\phi\right) = 2\pi$ in Eq. (1.20).

To find the function $f\left(\phi\right)$ in Eq. (1.20) for the pendulum, multiply both sides of Eq. (1.1) by $d\phi/dt$,

$$\frac{d^2\phi}{dt^2}\frac{d\phi}{dt} = \omega_0^2\frac{d\phi}{dt}\sin\varphi \qquad (1.21)$$

Integrating yields

$$\frac{1}{\omega_0}\frac{d\phi}{dt} = \sqrt{\cos\phi - \cos\phi_0} \qquad (1.22)$$

where ϕ_0 is the maximum value of the angle ϕ for which the angular velocity vanishes. Integrating again leads to

$$\int_0^\phi \frac{d\left(\phi/2\right)}{\left[\left(\sin^2\left(\phi/2\right) - \sin^2\left(\phi_0/2\right)\right)\right]^{1/2}} = \omega_0 t \qquad (1.23)$$

under the assumption that $\phi\left(t = 0\right) = 0$.

We introduce the variables ψ and k,

$$\sin\frac{\phi}{2} = k\sin\psi; \qquad k = \sin\frac{\phi_0}{2}. \qquad (1.24)$$

As the angle ϕ varies from 0 to ϕ_0, the angle ψ varies from 0 to $\pi/2$. Then, (1.23) becomes an elliptic integral of the first kind $F(k, \psi)$,

$$\omega_0 t = F\left(k, \psi\right); \qquad F\left(k, \psi\right) \equiv \int_0^\psi \frac{dz}{\sqrt{1 - k^2\sin^2 z}}. \qquad (1.25)$$

The time required for the rotation of the pendulum from $\phi = 0$ to $\phi = \phi_0$ equals one fourth the period T, which is given from (1.23) by $F\left(k, \pi/2\right)$, the complete elliptic integral of the first kind,

$$T = \frac{4}{\omega_0^2}F\left(k, \pi/2\right); \qquad F\left(k, \pi/2\right) \equiv \int_0^{\pi/2} \frac{dz}{\sqrt{1 - k^2\sin^2 z}}. \qquad (1.26)$$

Since $k < 1$, one can expand the square root in (1.26) in a series and perform a term-by-term integration,

$$T = \frac{2\pi}{\omega_0}\left[1 + \left(\frac{1}{2}\right)^2 k^2 + \left(\frac{1*3}{2*4}\right)^2 k^4 + ...\right]. \qquad (1.27)$$

Using the power-series expansion of $k = \sin\left(\phi_0/2\right)$, one can write a different series for T,

$$T = \frac{2\pi}{\omega_0}(1 + \frac{\phi_0^2}{16} + \frac{11\phi_0^4}{3072} + ...) \qquad (1.28)$$

The period of oscillation of the plane pendulum is seen to depend on the amplitude of oscillation ϕ_0. The isochronism found by Galileo occurs only for small oscillations when one can neglect all but the first term in (1.28).

To avoid elliptic integrals, one can use approximate methods to calculate the period T. For small ϕ, $\sin\phi \approx \phi$, and the linearized Eq. (1.1) describes the dynamics of a linear harmonic oscillator having solution $\phi = A\sin\left(\omega_0 t\right)$, with $\omega_0 = \sqrt{g/l}$. We use this solution as the first approximation to the nonlinear equation and insert it into (1.1). Repeating this process again and again, one obtains in first approximation,

$$\frac{d^2\phi}{dt^2} \approx -\omega_0^2 \left[A\sin(\omega_0 t) - \frac{[A\sin(\omega_0 t)]^3}{3!} + \frac{[A\sin(\omega_0 t)]^5}{5!} + ... \right] \quad (1.29)$$

Each term in (1.29) contains harmonics that correspond to a power of $A\sin(\omega_0 t)$, i.e., the series consists of the odd harmonics of the characteristic frequency ω_0 of the linear oscillator. The second approximation has a solution of the form $\phi = A\sin(\omega_0 t) + B\sin(3\omega_0 t)$. A complete description requires a full Fourier spectrum,

$$\varphi = \sum_{l=0}^{\infty} A_{2l+1} \sin[(2l+1)\omega_0 t] \quad (1.30)$$

Turning now to the calculation of the period T, one can use the following approximate method [7]. Since the period depends on the amplitude ϕ_0, one can write $T = T_0 f(\phi_0)$, where $T_0 = 2\pi\sqrt{g/l}$. One may then rewrite Eq. (1.1),

$$\frac{d^2\phi}{dt^2} + \frac{g}{l}\psi(\phi)\phi = 0, \quad \psi(\phi) = \left(\frac{\sin\phi}{\phi}\right) \quad (1.31)$$

and replace $\psi(\phi)$ by some average value $\psi(\overline{\phi})$. According to (1.24) and (1.27), $T = T_0(1 + \phi_0^2/16 + ...)$. Comparing the latter expression with the series expansion for $\psi(\overline{\phi})$ yields $\overline{\phi} = \sqrt{3}\phi_0/2$. Finally, one obtains the first correction T_1 to the period,

$$T_1 = T_0 \left(\frac{\sin(\sqrt{3}\phi_0/2)}{\sqrt{3}\phi_0/2} \right)^{-1/2} \quad (1.32)$$

A comparison between the approximate result (1.32) and the exact formula (1.26) shows that (1.32) is accurate to 1% for amplitudes up to 2.2 radian [7].

Another way to estimate the period of pendulum oscillations is by a scaling analysis [6]. In the domain $0 < t < T/4$, with characteristic angle $\phi(0) = \phi_0$, one obtains the following estimates

$$\frac{d\phi}{dt} \sim -4\frac{\phi_0}{T}; \quad \frac{d^2\phi}{dt^2} \sim -16\frac{\phi_0}{T^2}; \quad \sin\phi \sim \sin\phi_0 \quad \cos\phi \sim 1 \quad (1.33)$$

Inserting (1.33) into (1.1) yields

$$T\sqrt{\frac{g}{l}} \sim 4\left(\frac{\phi_0}{\sin\phi_0}\right)^{1/2} \qquad (1.34)$$

or

$$\frac{T}{T_0} \sim \frac{2}{\pi}\left(\frac{\phi_0}{\sin\phi_0}\right)^{1/2} \qquad (1.35)$$

Rewriting (1.22) as

$$\frac{1}{2}\left(\frac{d\phi}{dt}\right)^2 + \frac{g}{l}\left[\cos\phi_0 - \cos\phi\right] = 0 \qquad (1.36)$$

and inserting in (1.33) yields

$$\frac{T}{T_0} \sim \frac{2}{\pi}\frac{\phi_0/2}{\sin(\phi_0/2)} \qquad (1.37)$$

In the scaling analysis, one drops the numerical factor $2/\pi$ and, keeping the same functional dependencies, one writes the following general form of Eqs. (1.35) and (1.37),

$$\frac{T}{T_0} \approx \left[\frac{a\phi_0}{\sin(a\phi_0)}\right]^b \qquad (1.38)$$

Expanding $\sin(a\phi_0)$ yields

$$\frac{T}{T_0} = 1 + \frac{a^2 b}{6}\phi_0^2 + a^4 b\left(\frac{1}{180} + \frac{b}{72}\right)\phi_0^4 + \dots \qquad (1.39)$$

Comparing with (1.28) yields $a = 5\sqrt{2}/8$ and $b = 12/25$.

For completeness, we mention a few additional results [8]-[12] of approximate calculations of the period T of a pendulum as a function of the maximum angle ϕ_0

$$T_8 = T_0\left[\frac{1}{\cos(\phi_0/2 + \phi_0^3/256)}\right]^{1/2} \quad ; \quad T_9 = T_0\left[\frac{\phi_0}{\sin\phi_0}\right]^{3/8} \quad ; \qquad (1.40)$$

$$T_{10} = T_0\left[\frac{1}{\cos(\phi_0/2)}\right]^{1/2}$$

$$T_{11} = T_0\frac{\ln\cos(\phi_0/2)}{\cos(\phi_0/2) - 1} \quad ; \quad T_{12} = T_0\ln\frac{4}{\cos(\phi_0/2)} \qquad (1.41)$$

where the subindex of T shows the number of the appropriate Reference. Yet another method is based on the approximate calculation of the complete elliptic integral of the first kind [13].

1.3   Underdamped pendulum

The addition of damping to Eq. (1.1) makes it analytically unsolvable. Assuming that the damping is proportional to the angular velocity, the equation of motion becomes

$$\frac{d^2\phi}{dt^2} + \alpha\frac{d\phi}{dt} + \frac{g}{l}\sin\phi = 0 \qquad (1.42)$$

There are no chaotic solutions of Eq. (1.42). Almost all solutions of this equation describe phase space trajectories terminating at the stable fixed point $\phi = d\phi/dt = 0$, which "attracts" all trajectories from its "basin of attraction." Only very few initial conditions ("stable manifold") will lead to an unstable fixed point, $\phi = \pi$, $d\phi/dt = 0$. Moreover, the fixed point, $\phi = d\phi/dt = 0$, is linearly stable, i.e., small perturbations from this point will decay in time. However, the fixed point $\phi = \pi$, $d\phi/dt = 0$, is linearly unstable, which means that small perturbations from this point will grow exponentially in time. A distinction has to be made between the fixed points $\phi = \pm n\pi$, for odd n and even n. Indeed, for the stability analysis, it is enough to expand the nonlinear $\sin\phi$ term in a series near the fixed point $n\pi$, which gives $\sin(\phi - n\pi) \approx \pm(\phi - n\pi)$ for even and odd n, respectively. A simple stability analysis shows [14] that for even n, all trajectories will be attracted to the fixed point (focus), whereas for odd n, the trajectories terminate at these fixed points which are stable in one direction but unstable in other directions (saddle points).

Equation (1.42) does not have an analytical solution, and we content ourselves with numerical solutions. One proceeds as follows. Equation (1.42) can be rewritten as two first-order differential equations,

$$z = \frac{d\phi}{dt}; \quad \frac{dz}{dt} + \alpha z + \frac{g}{l}\sin\phi = 0 \qquad (1.43)$$

The $(d\phi/dt, \phi)$ phase-plane, shown in Fig. 1.3 [15], is different from that shown in Fig. 1.2 for the undamped pendula.

The higher derivatives $d^m\phi/dt^m$ are increasingly sensitive probes of the transient behavior and the transition from locked to running trajectories.

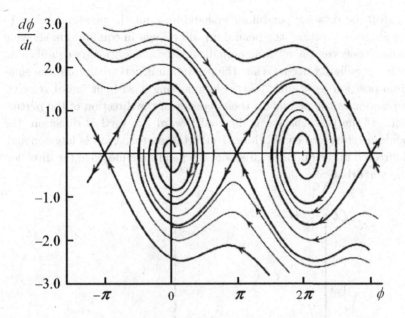

Fig. 1.3 Phase plane of a damped pendulum.

The higher derivatives of the solutions of Eqs. (1.43) are shown [16] in the $(d^m\phi/dt^m, d^{m-1}\phi/dt^{m-1})$ phase-planes for $m \leq 5$.

The pendulum equation with constant torque is obtained by adding a constant a to Eq. (1.42),

$$\frac{d^2\phi}{dt^2} + \alpha\frac{d\phi}{dt} + \frac{g}{l}\sin\phi = a \qquad (1.44)$$

This equation cannot be solved analytically, and we shall present the results of numerical calculations [17; 18].

For $a = 0$, the pendulum hangs in the downward position, $\phi = 0$. With increasing a, the equilibrium state is tilted to $\phi = \sin^{-1}(al/g)$, approaching $\phi = \pi/2$ at $a = g/l$. For $a > g/l$, the pendulum starts to rotate, corresponding to the loss of equilibrium. The type of rotation depends on the damping. If α is small, the pendulum rotates with a finite period. For strong damping and close to the threshold, $a - g/l$ of order the small parameter ε, and the period of rotation scales as $\varepsilon^{-1/2}$.

As a decreases from $a > g/l$ to $a < g/l$, the pendulum continues to rotate even for $a < g/l$, which is the manifestation of hysteresis. For $a \leq g/l$, if we stop the rotating pendulum by hand and put the pendulum close to its equilibrium position, the pendulum will remain in equilibrium, showing bistability between the rotating periodic regime and stationary equilibrium. Another peculiarity occurs when the pendulum is perturbed from the equilibrium position $\phi_1 = \sin^{-1}(a_1 l/g)$ by releasing it without initial velocity. The reaction of the pendulum then depends of the direction of the perturbation. If the perturbation is directed toward the vertical direction, the pendulum returns to equilibrium. But if the perturbation is large enough and directed in the opposite direction, the pendulum moves in the direction of the perturbation.

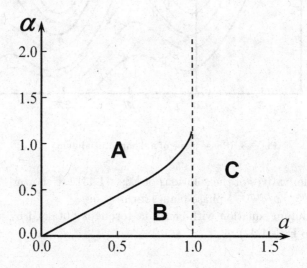

Fig. 1.4   Bifurcation diagram for a damped pendulum driven by a constant torque. Regions with stable solutions: A) Stationary, B) Periodic, C) Stationary and Periodic. Reprinted with permission from [17], Copyright (2005), American Association of Physics Teachers.

The results of the numerical solution of Eq. (1.44) are shown [17] for different types of stable solutions in the $\alpha - a$ phase diagram (Fig. 1.4). The three regions in this figure describe the stable solutions that are stationary (region A), periodic (region C) and both stationary and periodic (region B). There are two boundaries in this phase diagram: $a = 1$ which corresponds to the disappearance of equilibria, and another boundary which restricts the region where periodic trajectories exist. One can find analytically [17] the initial slope of the latter curve when both a and α are small, $a \sim \alpha \sim \varepsilon$, where $\varepsilon \ll 1$. Multiplying Eq. (1.44) by $d\phi/dt$ and integrating from t_i and t_f, one obtains the energy balance equation

$$\frac{1}{2}\left[\left(\frac{d\phi_f}{dt}\right)^2 - \left(\frac{d\phi_i}{dt}\right)^2\right] + \alpha \int_{t_i}^{t_f} \left(\frac{d\phi}{dt}\right)^2 dt - (\cos\phi_f - \cos\phi_i) = a\left(\phi_f - \phi_i\right) \tag{1.45}$$

The solution of this equation to order ε for $t_i = -\infty$, $t_f = \infty$, $\phi_i = -\pi$, $\phi_f = \pi$, $d\phi_i/dt = d\phi_f/dt = 0$ is

$$2\pi a = \alpha \int_{\phi_i}^{\phi_f} \left(\frac{d\phi}{dt}\right) dt = 8\alpha \tag{1.46}$$

Analogously to Eq. (1.43), Eq. (1.44) can be rewritten as two first-order differential equations by introducing the dimensionless angular velocity

$$z = \frac{\alpha l}{g}\frac{d\phi}{dt} \tag{1.47}$$

implying

$$\frac{dz}{dt} = \frac{gz}{\alpha l}\frac{dz}{d\phi} \tag{1.48}$$

Then, one can rewrite Eq. (1.44) in the form

$$\frac{bz}{\alpha^2}\frac{dz}{d\phi} + z + \sin\phi = \frac{al}{g} \tag{1.49}$$

where $al/g \equiv \bar{a}$ is the dimensionless torque.

The solutions of Eqs. (1.47)-(1.49) plotted in the $\langle z \rangle - \bar{a}$ plane are shown in Fig. 1.5. Two solutions exist for the average angular velocity $\langle z \rangle$ in the interval $1 > \bar{a} > \bar{a}_{cr}$, where $\bar{a}_{cr}$ depends on the parameter $g/l\alpha^2$. The smaller this parameter, the larger the bias torque $\bar{a}$ that is needed to start the motion, with some threshold value of $\langle z \rangle$. On the other hand, by decreasing the torque $\bar{a}$, the pendulum will continue to rotate until the torque reaches the value $\bar{a}_{cr}$, at which point the pendulum will come to rest after performing damped oscillations around the equilibrium position $\phi = \sin^{-1}\left(\bar{a}_{cr}/b\right)$.

The pendulum equations considered above are called the underdamped equations. In many cases, the first (inertial) term in (1.42) is small compared with the second (damping) term, and may be neglected. Redefining the variables, Eq. (1.42) reduces to the overdamped equation

$$\frac{d\phi}{dt} = a - b\sin\phi \tag{1.50}$$

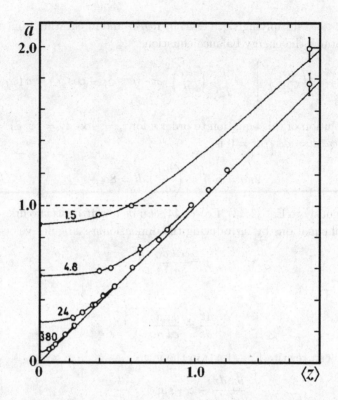

Fig. 1.5 Averaged angular velocity as a function of torque for different values of the parameter gl/α^2. Reprinted with permission from [18] , Copyright (1977), American Association of Physics Teachers.

The next complication of the original equation (1.1) involves introducing the driving force into Eq. (1.42),

$$\frac{d^2\phi}{dt^2} + \alpha\frac{d\phi}{dt} + \frac{g}{l}\sin\phi = A\sin(\Omega t) \qquad (1.51)$$

For this non-Hamiltonian "dissipative" system, the volume in phase space is not conserved. The trivial case is the contracting of all trajectories to the fixed point. A more interesting case is contracting the phase volume in one direction and stretching in the other direction, which leads to chaos.

1.4 Nonlinear vs linear equation

The nonlinear equation (1.51) has no analytic solution. However, for small oscillations, one can replace $\sin \phi$ by ϕ, and obtain the following equation

$$\frac{d^2\phi}{dt^2} + \alpha\frac{d\phi}{dt} + \frac{g}{l}\phi = A\sin(\Omega t) \qquad (1.52)$$

which can be solved analytically. For large t, the transient decays away, and the solution of Eq. (1.52) does not depend on the initial conditions, and has the following form,

$$\phi(t) = \frac{\alpha\Omega\cos(\Omega t) - (\Omega^2 - g/l)\sin(\Omega t)}{(\Omega^2 - g/l)^2 + \alpha^2\Omega^2} \qquad (1.53)$$

The response $\phi(t)$ to the driving force has a resonant character, being maximal at the driving frequency Ω. One way to represent the pendulum motion (1.53) geometrically is to use a phase diagram. For the second-order differential equation (1.52), the motion of a pendulum is specified by two parameters which define the $d\phi/dt$ - ϕ phase plane. Each point on this plane corresponds to a state of the pendulum at a given time t. Therefore, the trajectory is defined by Eq. (1.53) which corresponds to an ellipse,

$$\phi^2 + \frac{1}{\Omega^2}\left(\frac{d\phi}{dt}\right)^2 = \frac{A^2}{(g/l - \Omega^2)^2 + \alpha^2\Omega^2} \qquad (1.54)$$

For all initial conditions, the pendulum will finally reach this ellipse, which is therefore called the "periodic attractor." If there is no driving force ($A = 0$ in Eq. (1.52)), the pendulum will finally reach the downward position, which is called the "point attractor."

To compare the results obtained for a linearized pendulum with those for a nonlinear pendulum, one has to solve Eq. (1.51) numerically. Using the fixed driving amplitude and frequency, $A = 2$ and $\Omega = 2/3$, yields [14] the phase portraits for different values of the damping coefficient α. For large α, the pendulum executes small oscillations, which justify the approximation $\sin \phi \approx \phi$, and the resulting phase portrait is similar to that obtained for the linearized pendulum, slightly increasing due to an increase of oscillation. However, unexpected changes occur when the parameter α decreases further [14]. First, the ellipse changes its form and shifts to the left displaying "symmetry breaking" with respect to the vertical axis (note that there are no non-symmetric terms in Eq. (1.51)). Second, further decrease of the parameter α leads to the appearance of a second ellipse

("period-doubling bifurcation"), i.e., the appearance of two periodic attractors, then additional bifurcation to four attractors, etc. Depending on
initial conditions ("basin of attraction"), a pendulum will reach one of possible attractors. Finally, the accumulation points leads to chaotic behavior,
the most important characteristic of which is the exponential separation
of trajectories for very small differences in initial conditions. The "chaotic
attractor" possess different properties which we will consider in Section 2.2.

1.5   Isomorphic models

In the examples given below, for simplicity we will compare the derived
equation with the overdamped pendulum Eq. (1.50), whereas their more
complete description is equivalent to the underdamped pendulum Eq.
(1.42).

1.5.1   *Brownian motion in a periodic potential*

Replacing the angular variable ϕ in Eq. (1.1) by the coordinate x, yields
the equation describing one-dimensional motion of a Brownian particle in
a periodic potential. The literature on this subject is quite extensive (see,
for example, an entire chapter in Risken's monograph [19]).

1.5.2   *Josephson junction*

A Josephson junction consist of two weakly coupled superconductors, separated by a very thin insulating barrier. Since the size of the Cooper
pair in superconductors is quite large, the pair is able to jump across the
barrier to produce a current, the so-called Josephson current. The basic equations governing the dynamics of the Josephson effect connects the
voltage $U(t) = (\hbar/2e)(\partial\phi/\partial t)$ and the current $I(t) = I_c \sin\phi(t)$ across
the Josephson junction. This defines the "phase difference" ϕ of the two
superconductors across the junction. The critical current I_c is an important phenomenological parameter of the device and it can be affected by
temperature as well as by an applied magnetic field. The application of
Kirchoff's law to this closed circuit yields

$$I = I_c \sin\phi(t) + \frac{\hbar}{2eR}\frac{\partial\phi}{\partial t} \qquad (1.55)$$

where R is the resistivity of the circuit, and I and I_c are the bias and critical current, respectively. This equation is simply Eq. (1.50).

1.5.3 *Fluxon motion in superconductors*

The magnetic field penetrates type-II superconductors in the form of quasi-particles called fluxons. In many cases, fluxons move in a periodic potential which is created by the periodic structure of pinning centers or by the plane layers of the superconductor. If one neglects the fluxon mass, the equation of motion of a fluxon has the form (1.50) [20].

1.5.4 *Charge density waves*

As a rule, in solids at low temperatures, the electron charge density is distributed uniformly. The well-known violation of this rule occurs in superconducting materials where the electrons are paired. Another example of the non-uniform distribution of electrons is the charge density wave (CDW), which behaves as a single massive particle positioned at its center of mass. CDWs have a huge dielectric constant, more than one million times larger than that of ordinary materials. One can clearly see the inhomogeneity of a CDW by the scanning tunneling microscope. Such a system shows "self-organization" in the sense that a small perturbation is able to induce a sudden motion of the entire charge density wave. Such a perturbation can be produced by an external electric field whereby an increase in voltage beyond a certain threshold value causes the entire wave to move, producing "non-Ohmic" current which vastly increases with only a small increase in voltage.

The mathematical description of the CDW based on the "single-particle" model assumes that the CDW behaves as a classical particle. The experimentally observed nonlinear conductivity and the appearance of a new periodicity together form the basis for the periodically modulated lattice potential acting on the CDW. The equation of motion of the center of mass x of a damped CDW has the following form,

$$\frac{d^2x}{dt^2} + \gamma \frac{dx}{dt} + b \sin x = 0. \tag{1.56}$$

This equation is the same as Eq. (1.42) for a damped pendulum.

1.5.5   *Laser gyroscope*

A mechanical gyroscope with rotating wheels is widely used for orientation in space. Nowadays, however, this is being replaced by the laser gyroscope which works on a physical principle established by Sagnac about a hundred years ago. Sagnac found that the difference in time for two beams traveling in opposite directions around a closed path going through a rotating platform is proportional to the speed of the platform. The beam travelling in the direction of rotation of the platform travels for a longer distance than the counterrotating beam, and hence has a lower frequency. The phase difference ϕ between these two beams running in a ring-laser microscope allows one to find the velocity of the rotating platform. Equation (1.50) yields ϕ, where a denotes the rotation rate and b is the backscattering coefficient.

1.5.6   *Synchronization phenomena*

In the 17th century, the Dutch physicist Christian Huygens found that two pendulum clocks attached to a wall always run at the same rate, because the wall introduces a weak coupling between them. This phenomenon of synchronization is present in general in dynamic systems with two competing frequencies. The two frequencies may arise through the coupling of an oscillator to an external periodic force. The equation which describes the influence of a small external force on the intrinsic periodic oscillations of an oscillator [21] connects the phase difference ϕ between the oscillator frequency and that of an external force expressed by the frequency difference a, and the periodic force $b \sin \phi$. It has the form of Eq. (1.50) [22].

1.5.7   *Parametric resonance in anisotropic systems*

The rotation of an anisotropic cluster in an external field is described by the following equation,

$$\frac{d\mathbf{L}}{dt} = \mathbf{M} \times \mathbf{F} - \beta \boldsymbol{\omega} \qquad (1.57)$$

where $\mathbf{L}$ is the angular momentum, $\boldsymbol{\omega}$ is the angular velocity of rotation, $M_i = \chi_i F_i$ is the magnetic (dielectric) moment induced by the external field $\mathbf{F}$. Anisotropy means that $\chi_3 \neq \chi_1 = \chi_2$; $\Delta \chi \equiv \chi_1 - \chi_3$, while the moment of inertia is isotropic in the $x - y$ plane, $I_1 = I_2 \equiv I$. Connecting the coordinate axis with the moving cluster, one can easily show [23] that the equation of motion for the nutation angle θ coincides with Eq. (1.42) for a damped pendulum,

$$\frac{d^2\theta}{dt^2} + \frac{\beta}{I}\frac{d\theta}{dt} + \frac{F^2\Delta\chi}{2I}\sin(2\theta) = 0. \tag{1.58}$$

For an alternating external field, $F = F_0\cos(\omega t)$, Eq. (1.58) takes the form of the equation of motion of a pendulum with a vertically oscillating suspension point (see Section 2.3.1). Absorption of energy by clusters has a resonant character: only clusters of specific site are able to rotate in the external field of given frequency and amplitude. This new type of parametric resonance in magnetically (electrically) anisotropic systems can be used for an experimental investigation of polymers, liquid crystals, viruses, proteins and the kinetics of liquid-solid transitions.

1.5.8 *Phase-locked loop*

A phase-locked loop (FLL) is a closed-loop frequency-control electronic circuit with a voltage - or current-driven oscillator adjusted to match in phase (and thus lock on) the frequency of an input signal. The FLL is also described by the pendulum equation (1.42) [24]. The period-doubling transition to chaos in FLL was experimentally observed [25] by the appropriate change of control parameter. FLL is used in radio and TV sets to obtain stable tuning [19].

1.5.9 *Dynamics of adatom subject to a time-periodic force*

The motion of atomic and molecular adsorbates on a metallic or semiconductor surface is described by the pendulum equation (1.42) [26]. The theoretical calculations are important in connection with modern experimental techniques, such as field ion or scanning tunneling microscopy and quasi-elastic helium-atom scattering, allowing the precise measurement of adatoms. Many references can be found in [27].

1.5.10 *The Frenkel-Kontorova model (FK)*

In its simplest form, the FK model describes the motion of a chain of interacting particles ("atoms") subject to an external on-site periodic potential [28]. This process is modulated by the one-dimensional motion of quasi-particles (kinks, breathers, etc.). The FK model was originally suggested for a nonlinear chain to describe, in a simple way, the structure and dynamics of a crystal lattice in the vicinity of a dislocation core. It was subsequently used to describe different defects, monolayer films, proton conductivity of hydrogen-bonded chains, DNA dynamics and denaturation.

1.5.11 Solitons in optical lattices

Although solitons are generally described by the sine-Gordon equation, the motion of the soliton beam in a medium with a harmonic profile of refractive index is described by the pendulum equation with the incident angle being the control parameter [29].

1.5.12 Other applications

Some unexpected applications of the pendulum equations include the problem of energy extraction from ocean waves [30], gravitational gradient pendulum [31] and ship dynamics [32], among others. Less serious applications include the description of toys, such as the space circle [33] and tilt-a-wirl [34].

1.6 General concepts

Deterministic chaos results if the differential equations are nonlinear and contain at least three variables. This points to the important difference between the underdamped and the overdamped equations of motion of a pendulum. The underdamped equation, subject to an external periodic force,

$$\frac{d^2\phi}{dt^2} + \gamma\frac{d\phi}{dt} + \sin\phi = f\sin(\omega t),\qquad(1.59)$$

can be rewritten as a system of three first-order differential equations,

$$\frac{d\chi}{dt} + \gamma\chi + \sin\phi = f\sin(\theta)\qquad(1.60)$$

$$\frac{d\phi}{dt} = \chi;\qquad\frac{d\theta}{dt} = \omega\qquad(1.61)$$

Therefore, the underdamped equation exhibits deterministic chaos for some values of the parameters. On the other hand, the overdamped equation, in which $d^2\phi/dt^2 = 0$, has only two variables, and therefore, it does not exhibit chaos. We consider here only the noise-free underdamped pendulum, starting from Eq. (1.59).

There are many books describing deterministic chaos, including [14], which considers chaos for the damped driven pendulum. We shall present only those basic concepts which are needed in the later discussion.

1.6.1 *Phase space*

The motion of a pendulum is conveniently displayed graphically in the plane of its phase variable, angle ϕ and angular velocity $d\phi/dt$. The characteristic points or curves in these diagrams are called "attractors" because irrespective of initial conditions, all trajectories are asymptotically attracted to them. If the pendulum remains at a stable downward position, the origin will be the "point attractor," whereas for periodic motion, the "periodic attractor" will have the form of an ellipse.

1.6.2 *Poincare sections and strange attractors*

The Poincare section is obtained by removing one space dimension. In the case of three variables, the Poincare section is a plane of $d\phi/dt$ - ϕ variables. As distinguished from the $d\phi/dt$ - ϕ phase portrait considered in Section 1.1, which shows the entire pendulum trajectory, the Poincare section gives the stroboscopic section of this trajectory. The Poincare plane has to be chosen in such a way that trajectories will intersect it several times. If the motion is periodic (non-chaotic), the trajectory will cross the Poincare plane repeatedly at the same point. The number of points appearing in the Poincare plane defines the number of periods corresponding to a given trajectory. However, chaotic motion means that despite the fact that the motion is deterministic (i.e., for given initial conditions, the equations exactly describe the trajectory), it never repeats itself. However, there will be a dense set of points in the Poincare section filling a certain area of this plane. The locus of these points, which is a wavy line, is called a "strange attractor," and the set of all initial points that eventually bring the system to these attractors are called the basin of the attractor. In this way, one reduces the continuous trajectories in the phase space to a discrete mapping on the Poincare plane. The strange attractor that contains many lines, has a fractal structure, which means that upon magnifying one of these lines, one sees a group of similar lines, etc. In summary, the regular orbits appear as points of a smooth area whereas the chaotic trajectories appear as a platter of points filling a certain area.

1.6.3 *Lyapunov exponent*

The Lyapunov exponent λ gives the rate at which the two nearby trajectories, separated initially by distance d, become separated in time by a very large distance L

$$L = d \exp(\lambda t) \tag{1.62}$$

In a chaotic regime, λ is positive, whereas in the regular regime, λ is negative or zero, implying that the initial separation either diminishes or remain constant. For a system of first-order differential equations, there are several Lyapunov exponents, and we are interested in the largest of these.

1.6.4 *Correlation function*

The analysis of the autocorrelation function of a trajectory shows the difference between regular and chaotic regimes. In the latter case, the system loses information about previous states and the autocorrelation function tends to zero with time. However, since chaotic trajectories densely fill phase space, there is some short time during which the trajectory approaches the initial position, and the autocorrelation function may grow. In contrast to these results, for the regular trajectory, the autocorrelation function oscillates around some average value, increasing and decreasing as the system moves away or approaches its initial position. The difference in the form of the autocorrelation function for regular and chaotic regimes is shown in Fig. 1.6 [35].

1.6.5 *Spectral analysis*

The squared modulus of the Fourier transform of the variable $g(t)$ is called the power spectrum of $g(t)$. Periodic motion is described by a power spectrum with a finite number of frequencies, whereas chaotic motion has a broad power spectrum without any well-defined peaks. In Figs. 1.7 and 1.8, we show [36] the efficiency of the four methods of detecting chaos described above, using Eq. (1.59) as an example. A comparison between Figs. 1.7 and 1.8 shows that although some methods (power spectrum) are ambiguous, one can see a clear distinction between regular (Fig. 1.7) and chaotic (Fig. 1.8) behavior.

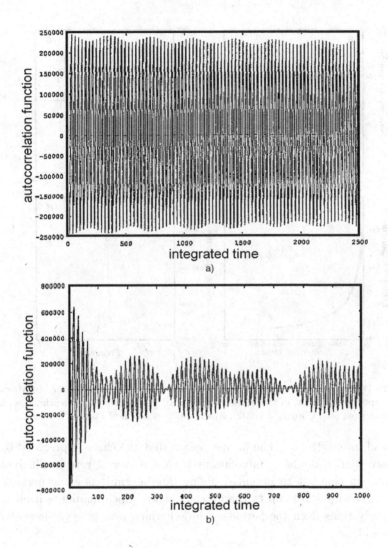

Fig. 1.6 Autocorrelation functions for a) regular region with energy $E = 0.00375$, and b) chaotic region with $E = 0.04875$. Reprinted with permission from [35], Copyright (1994), Institute of Physics.

1.6.6 *Period doubling and intermittency*

Starting from the periodic regime, the change of some parameters in the dynamic equations may move the system into the chaotic regime [37]. The most common ways of entering the chaotic regime are through period dou-

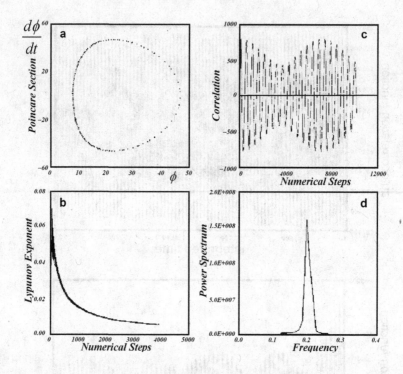

Fig. 1.7   Different indications for a regular trajectory: a) Poincare section, b) Maximum Lyapunov exponent, c) Correlation function, d) Power spectrum. Reprinted with permission from [36], Copyright (1992), American Association of Physics Teachers.

bling and intermittency. The former means that the chaos is preceded by successive period-doubling bifurcations with a universal rate in different systems. In contrast to period doubling, for intermittency the periodic solution remains stable up to the onset of chaos which manifests itself in short derivations from the periodic solution which take place at irregular intervals [38].

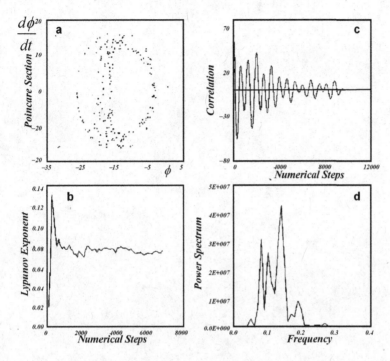

Fig. 1.8   Same as Fig. 1.7. but for a chaotic trajectory. Reprinted with permission from [36], Copyright (1992), American Association of Physics Teachers.

Chapter 2

Deterministic Chaos

2.1 Damped, periodically driven pendulum

Equation (1.59) can be rewritten as

$$\frac{d^2\phi}{dt^2} + \alpha\frac{d\phi}{dt} + \sin\phi = A\sin(\Omega t) \tag{2.1}$$

where the applied frequency Ω has been normalized to the (linearized) pendulum frequency ω_0, and the time t is normalized to ω_0^{-1}. Equation (2.1) contains three independent parameters: the strength of the damping α, and the amplitude A and frequency Ω of the external field. After considering the general properties of Eq. (2.1) in Section 2.1.1, we analyze the influence of a second periodic field in Section 2.1.2, and the influence of parameters Ω and α in Sections 2.1.3 and 2.1.4. Finally, Sections 2.1.5 and 2.1.6 are devoted to symmetry changes and anomalous diffusion.

2.1.1 Transition to chaos

Although Eq. (2.1) does not have an analytical solution, qualitative comments can be made [39]. As previously stated, the existence of chaotic solutions of a given differential equation requires two conditions. First, the equation must be nonlinear, which is provided by the $\sin\phi$ term in Eq. (2.1). Second, the equation must have at least three dynamic variables, which also applies to Eq. (2.1), as was shown in Eqs. (1.60) and (1.61). Since the nonlinearity and minimum dimension three of a differential equation are the necessary conditions for chaotic behavior, no chaotic solutions will exist for a small driving amplitude A and large damping coefficient α. Indeed, in the former case, the system oscillates in the small amplitude regime where it is linear, while in the latter case, the differential equation

27

system has only dimension two. Chaotic behavior exists for the driving frequency close to the resonant eigenfrequency of a (linearized) pendulum, i.e., chaotic effects will occur near $\Omega \approx 1$.

The general properties of the solutions of Eq. (2.1) can be inferred from the comparison of the driving amplitude A and the magnitude of the torque acting on the pendulum, which is equal (in dimensionless units) to $\sin \phi$, i.e., equal to unity for $\phi = \pi/2$. For a weak external force, $A < 1$, the pendulum will remain at $\phi < \pi/2$, executing oscillations near the downward position. However, for $A > 1$, the strong driving force will lead to full rotations,

For different values of the parameters, Eq. (2.1) exhibits a great variety of different solutions (running and locked, symmetric solutions to the downward positions and symmetry-breaking solutions which oscillate with a larger amplitude to one side than to the other, and, finally, chaotic solutions). Three different paths to chaos have been found [40]: 1) Period-doubling cascade preceded by the appearance of symmetry-breaking solutions. This cascade is produced for both running and locked solutions. In the latter case, the frequency is locked to the external frequency. 2) The loss of phase locking and random transitions between two locked states. 3) The intermittency form of the transition to chaos occurring for an external field of large amplitude (larger than the amplitude defining the transition from locked to running solutions). The trajectories are then a combination of clockwise and counter-clockwise rotations with damped oscillations in-between.

The theoretical explanation of the observed striking noise-rise phenomena in a Josephson junction was first presented in a pioneering work [41]. This explanation turns out to be a special case of the general phenomenon of "deterministic chaos" in nonlinear dynamics. By means of a hybrid digital-analog computer system, Eq. (2.1) has been solved for $\alpha = 0.5$ by varying Ω at constant A. The bifurcation diagram (Fig. 2.1) shows the different types of behavior for different values of driving amplitude and frequency.

This graph shows the variety of different types of dynamic behavior of a system described by Eq. (2.1). For $\Omega \gg 1$ or $\Omega \ll 1$ (the external frequency is much larger or much smaller than the oscillator frequency), one obtains periodic solutions, which can be quite complicated with sub-harmonics, harmonics, hysteresis loops, etc. For different regions of the $A - \Omega$ plane, one obtains locked or running solutions. At large amplitudes A, the pendulum performs rotations corresponding to the periodic motion from one potential well to another (region A). In region B, one encounters oscillations near the downward position. In this region, the set of period-

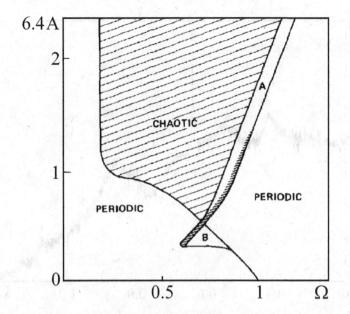

Fig. 2.1 Bifurcation diagram (normalized amplitude as a function of frequency) for a periodically driven damped pendulum for $\alpha = 0.5$. Reprinted with permission from [41], Copyright (1980), American Institute of Physics.

doubling bifurcations appears, leading to a chaotic state. Here the angle ϕ is no longer bound, and the turbulent behavior is characterized by strange attractors. The structure of these strange attractors can be explained [41] in terms of two distinct time scales of the motion: the shorter time corresponds to fast oscillations between closely related wells, whereas the longer time scale is associated with a slower diffusion between separated wells. This chaotic motion, generated by the deterministic Eq. (2.1), is characterized by the power spectral density $S(\omega)$ of the autocorrelation function for the time derivative $d\phi/dt$ shown in Fig. 2.2 for the parameters $A = 3.8$ and $\Omega = 0.64$. The large peak observed at the driving frequency broadens at larger driving amplitudes.

The solutions of Eq. (2.1) have been obtained [42] on a specially constructed experimental setup for fixed values of the damping constant α and the driving frequency Ω : $\alpha = 0.24$, $\Omega = 0.67$. For small values of the driving amplitude A, the motion is periodic — the pendulum oscillates back and force and the spectrum of $\phi = \phi(t)$ has a principal peak at the driving frequency Ω and smaller peaks at the harmonics of Ω. For slightly larger

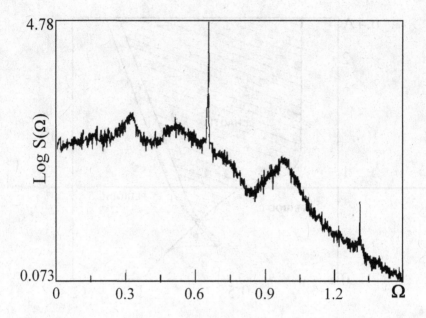

Fig. 2.2   Power spectrum of the strange attractors for a periodically driven damped pendulum with $A = 3.8$, $\Omega = 0.64$ and $\alpha = 0.5$. Reprinted with permission from [41], Copyright (1980), American Institute of Physics.

amplitude A, the noise background increases, and for $A = 0.69$, a period-two orbit appears. The power spectrum now contains peaks at integer multiples of $\Omega/2$. Another period-doubling then occurs, and chaos develops fully at $A \approx 0.70$.

The onset of chaos in a Josephson junction, described by Eq. (2.1), is connected [43] with a phenomenon called "high parametric gain" (stochastic amplification of a weak harmonic signal), manifested as a maximum of $d\phi/dt$ (voltage in a Josephson junction) as a function of Ω. Note that for different values of the parameters, this large gain appears in the absence of chaos. As is clear from the numerical calculations performed for $\alpha = 0.2$, both these phenomena appear for the same values of the parameters. This result was supported by the appearance of strange attractors on the Poincare map and by analysis of the graph for $\phi(t)$, as well as the Fourier spectrum of $d\phi/dt$ [43].

An unexpected application has been suggested [44] for the order-chaos transition for generating a high level of noise which is approximately white over a broad frequency region. The idea is based on the fact that the main property of white noise — absence of correlation between times t_1 and t_2 — which applies in the case considered here of motion in an exact periodic potential, allows the particle to forget its location and limits the time over which the correlation occurs. The calculated power spectrum of $d\phi/dt$ for Eq. (2.1) for parameters $\alpha = 0.1$, $A = 1.6$ and $\Omega = 0.8$ is displayed in Fig. 2.3, which shows that the "white" portion of this spectrum extends over four decades of frequency [44].

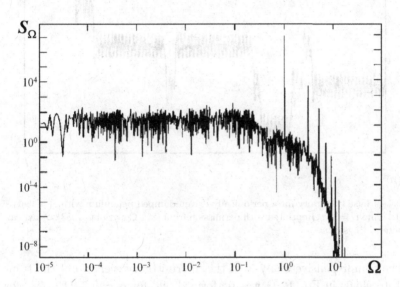

Fig. 2.3 Power spectrum of the angular velocity for a periodically driven damped pendulum with $A = 1.6$, 60, $\Omega = 0.8$, and $\alpha = 0.1$. Reprinted with permission from [44], Copyright (1999), American Institute of Physics.

A different intermittent (or tangent bifurcation) transition to chaos was found [45] from the numerical and analog solutions of Eq. (2.1) by changing the amplitude A for fixed $\alpha = 0.5$ and $\Omega = 0.47$. There exists a critical value of A, denoted A_C (in the language of the Josephson junction, $A_C = 0.875 \ mA$), such that for $A < A_C$, one obtains stable oscillations having the driving frequency. For A slightly larger than A_C, the system enters the chaotic regime via an intermittent transition as A is decreased. At this transition, $d\phi/dt$ becomes unstable and periodic oscillations occur, randomly interrupted by bursts of $d\phi/dt$. These bursts are shown in Fig. 2.4, where ϕ is plotted as function of time t [45]. The experimental results agree with theoretical predictions [38].

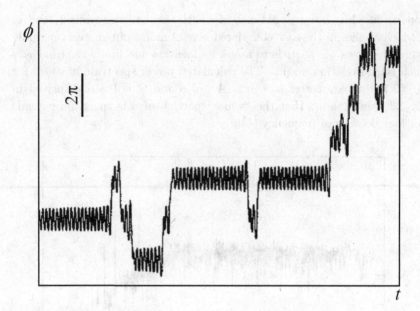

Fig. 2.4   Phase trajectory for a periodically driven damped pendulum with $A = 09045$, $\Omega = 0.47$ and $\alpha = 0.5$. Reprinted with permission from [45], Copyright (1983), American Institute of Physics.

The comprehensive analysis of the intermittency scenario of the transition to chaos in Eq. (2.1) was performed [46] for $\alpha = 0.5$, $\Omega = 2/3$ for different values of the amplitude A of an external field. For $A \leq 1.5$, there exist two separate stable periodic running solutions, clockwise and counterclockwise. As A increases, these two modes remain separate while becoming chaotic. At the critical value $A = 1.4945$, intermittent switching between these two modes occurs, producing a large amount of noise at frequencies smaller than the driving frequency. These results have been confirmed by analyzing the $(\phi, d\phi/dt)$ phase portrait and the Poincare sections [47].

2.1.2   *Two external periodic fields*

As seen in the previous section, a pendulum subject to an external periodic field (Eq. (2.1)) shows chaotic behavior for some values of the control parameters. One can change this behavior and pass to deterministic trajectories by changing the values of the parameters. Another way for taming chaotic dynamics is by adding a weak periodic perturbation. Consider the

following equation

$$\frac{d^2\phi}{dt^2} + \alpha\frac{d\phi}{dt} + \sin\phi = A_1\sin(\Omega_1 t) + A_2\sin(\Omega_2 t) \qquad (2.2)$$

Numerical calculations have been performed [48] for $\alpha = 0.7$ and $\Omega_1 = 0.25$. For $A_1 = 1.75$ and $A_2 = 0$, as well as for $\Omega_2 = 0.1$ and $A_2 = 0.0125 \ll A_1$, the behavior of the system is chaotic. However, for larger values of the second field, $A_2 = 1$, the motion is no longer chaotic, being replaced by periodic rotations.

Calculations have also been performed for an additional external constant force in Eq. (2.2) (ac current for Josephson junction) having the following equation of motion

$$\frac{d^2\phi}{dt^2} + \alpha\frac{d\phi}{dt} + \sin\phi = a + A_1\sin(\Omega_1 t) + A_2\sin(\Omega_2 t) \qquad (2.3)$$

When $A_2 = 0$ and $a = 0.905$, the solution of Eq. (2.3) is chaotic with the maximum Lyapunov exponent equal to about 0.04 [49]. However, by adding a small perturbation with $A_2 = 0.0125$, the situation is changed drastically. Figure 2.5 shows the maximum Lyapunov exponent λ as a function of Ω_2/Ω_1.

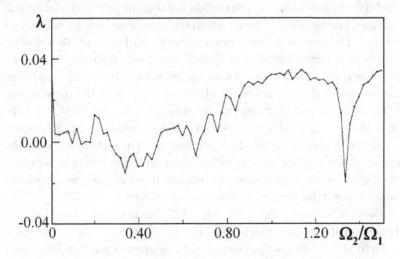

Fig. 2.5 Maximum Lyapunov exponent as a function of the ratio of two frequencies for a damped pendulum subject to constant torque a and two periodic forces with $a = 0.905$, $\alpha = 0.7$, $A_1 = 0.4$, $A_2 = 0.0125$. Reprinted with permission from [49]. Copyright (1991) by the American Physical Society.

One observes a significant reduction of λ, as well as several regimes of Ω_2/Ω_1 for which λ is negative, i.e., the motion is not chaotic. The sensitivity of chaotic dynamics to perturbations of initial conditions is complimented by its strong dependence on weak time-dependent perturbations. These results are supported quantitatively by experiments performed on an electronic Josephson junction simulator driven by two independent fields which also show the quasiperiodicity-chaos transitions induced by a weak external field [50], which is different from the transition to chaos through period-doubling bifurcations and intermittency.

2.1.3 *Dependence on driving frequency*

There are three parameters in Eq. (2.1) having the dimension of inverse time: the (small oscillation) pendulum frequency ω_0, which is equal to unity in dimensionless Eq. (2.1), the driving frequency Ω, and the damping coefficient α. For chaotic trajectories, Ω has to be smaller than ω_0, that is, $\Omega < 1$, since Ω has to be close to the characteristic frequency of the pendulum which becomes smaller than ω_0 due to nonlinear effects. The comprehensive numerical analysis of Eq. (2.1), performed for $\Omega < 1$, shows [40] that for the driving amplitude A larger than the critical value at which the pendulum begins to rotate, the motion is a combination of clockwise and anti-clockwise rotations in-between which the pendulum undergoes damped oscillations. The ensuing sensitivity on initial conditions leads to a chaotic state. There is some difficulty in finding the minimal driving frequency $\Omega_{\min}$ at which chaos still exists. In an earlier work [41], it was found that the frequency cutoff $\Omega_{\min}$ occurs at about α, whereas in the more recent work [40], it was found that the chaotic region is extended to $\Omega \ll 1$. In the limit of $\Omega \ll \alpha$, (the driving period is much longer than the damping time of the oscillations), as the driving amplitude passes through zero, the pendulum almost reaches its equilibrium position before the next rotation. Interestingly, a chaotic state exists in this low-frequency regime where the pendulum follows the excitation almost adiabatically.

For the opposite case of $\Omega \gg 1$, Eq. (2.1) has been solved numerically and (approximately) analytically [51]. For fixed values of the parameters, $\alpha = 0.2$ and $\Omega = 2$, the oscillations around the upward position have been found and exhibit no chaotic regime far from the resonance condition $\Omega = 1$.

The conventional stationary solution of Eq. (2.1) is

$$\phi(t) = A_1 \cos\theta_1 + A_3 \cos\theta_3 + A_5 \cos\theta_5 + ...; \quad \theta_m = m\omega t - \alpha_m; \quad 0 < \alpha_m < \pi \tag{2.4}$$

The coefficient A_1 increases with A, and when A exceeds ≈ 8.77, A_1 exceeds $A_c \approx 2.425$, which is close to the first zero (2.405) of the Bessel function $J_0(A_1)$. Symmetry breaking then occurs with respect to $\phi \to -\phi$, and the stationary solution (2.4) is replaced by the symmetry-breaking solution

$$\phi(t) = A_0 + A_1 \cos\theta_1 + A_2 \cos\theta_2 + A_3 \cos\theta_3 + ... \qquad (2.5)$$

At this stage, deterministic chaos-like phenomena occur. A change in the initial condition $\phi(t = 0)$ as small as 10^{-7} leads to a change of A_0 in (2.5) by a significant amount. However, such sensitivity to the initial conditions exists only for particular values of A, in contrast to deterministic chaos where sensitivity exists almost everywhere throughout a definite region of parameter space. On further increase of A, $|A_0|$ reaches π at $A = 10.58$, where the pendulum oscillates around the upward position. As A continuous to increase, A_1 increases while $|A_0|$ remains at π until A_1 reaches approximately the second zero (5.43) of $J_0(A_1)$. Then, $|A_0|$ decreases with A_1 until A_0 reaches zero and the pendulum returns to the stationary regime (2.4). Then, A_1 again increases until it reaches the third zero (8.65) of $J_0(A_1)$, and so on. In the regime $A \in (3.16, 5.97)$, a period-triple state appears, approximately described by $\phi = A_1 \cos(\omega t - \alpha_1) + A_{1/3} \cos[(\omega/3)t - \alpha_{1/3}]$. In the regime $A \in (3.77, 11.40)$, a sinusoidally modulated rotational state appears, approximately described by $\phi = A_0 + A_1 \cos(\omega t - \alpha_1) \pm \omega t$. The two signs correspond to the two directions of rotation. Note that these two modulated, period-triple states, along with the single-period states (2.4) and (2.5), can occur for the same value of A, with the initial conditions determining which state is actually excited. All these results obtained from numerical calculations are in good agreement with the analytic calculations performed in the framework of perturbation theory [51].

2.1.4 *Role of damping*

The equation of motion (2.1) describes not only the dynamics of the pendulum, but also such diverse phenomena as Josephson junctions and charge-density waves with low and high dissipation. The role of dissipation is of interest as a part of the general problem of nonlinear dynamics. Numerical solutions of Eq. (2.1) have been obtained for several values of α and, in each case, for a large number of values of A and Ω using a predictor-corrector method. The full $A-\Omega$ phase diagram has been obtained for small damping, $\alpha = 0.5$, which shows both period-doubling and the intermittent transition to chaos. These transient regions are very narrow. For example, for $\Omega = 0.09$, the entire period-doubling cascade occurs over a range $A_\infty - A_2 \approx 0.085$. Moreover, on the low-regime side, chaotic states adjoin the periodic state without a transition region, depending on the symmetry properties which we will discuss in the next section. Such direct transitions have also been obtained [52] for $\alpha = 0.2$, $\Omega = 0.6$, and $A = 0.61279$. At high dissipation, no chaotic behavior is found, which might be due to the fact that such system are described by the overdamped version of equation (2.1).

In addition to linear damping, some versions of the nonlinear damped pendulum have also been analyzed. The damping term, $\alpha \, d\phi/dt$ in Eq. (2.1), was replaced by $(1 + \gamma \cos \phi) \, d\phi/dt$ [53]. The transition to chaos was studied for parameters $\alpha = 0.4$, $A = 0.8$, $\gamma = -0.8$ and decreasing values of the driving frequencies Ω. For $1.2 \geq \Omega \geq 0.718$, only one symmetric solution exists. From $\Omega \approx 0.718$, a symmetry-breaking bifurcation takes place, leaving two symmetry-breaking solutions symmetric with respect to the origin on the phase plane diagram. The first period-doubling occurs at $\Omega \approx 0.685$ which, after additional period-doublings, leads to two independent chaotic attractors. For Ω varying from 0.73 to 0.45, alternative intervals of periodic and chaotic behavior occur. Two interesting sudden changes in chaotic behavior ("crises") occur at $\Omega \approx 0.67$ and $\Omega \approx 0.55$, where a large chaotic attractor is created from two independent chaotic attractors or from these attractors and two distinct unstable orbits. A comprehensive analysis of the phase diagram and power spectrum has also been performed for quadratic [54] and cubic [55] damping.

2.1.5 *Symmetry and chaos*

As noted previously [47], Eq. (2.1) is invariant under a simultaneous inversion of the pendulum angle ϕ and shift in phase of the driving force by an odd multiple of π,

$$\phi \to -\phi; \qquad t \to t + (2n+1)T/2 \qquad (2.6)$$

where T is the period of the driving force and n is an integer. Therefore, if $\phi(t)$ is a solution of Eq. (2.1), then $-\phi(t + (2n+1)T/2)$ is also a solution. The question arises whether these two functions are the same. If they are essentially the same, then the solution is said to be *symmetric*. "Essentially the same" means that they can differ by an integer number of complete revolutions,

$$-\phi(t + (2n+1)\,T/2) = \phi(t) + 2\pi n \qquad (2.7)$$

Otherwise, both solutions are termed *broken-symmetric*. An interesting connection exists [52] between this symmetry property and the development of chaotic solutions via the period-doubling mechanism. It follows from (2.7) that

$$\frac{d}{dt}\left[\phi(t + (2n+1)\,T/2)\right] = -\frac{d\phi}{dt} \qquad (2.8)$$

which means that a symmetric solution is a periodic solution with a period $T_s = (2n+1)\,T$, which is an odd-integer multiple of the driving period T. Averaging Eq. (2.8) over T_s shows that $\langle d\phi/dt \rangle = 0$, and the even-integer multiples of the fundamental frequency $\omega_s = 2\pi/T_s$ are missing in the power spectrum of the symmetric solutions of Eq. (2.1).

The numerical solution of Eq. (2.1) shows the change in symmetry upon the monotonic change of the control parameters [56]. Figure 2.6 shows the change of the angular velocity $d\phi/dt$ of the Poincare section as a function

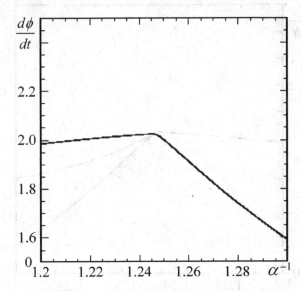

Fig. 2.6 Angular velocity as a function of the inverse damping parameter for a periodically driven damped pendulum with $A = 1.5$, $\Omega = 0.67$ and initial conditions $\phi_0 = (d\phi/dt)_0 = 0$. Reprinted with permission from [56]. Copyright (2006) Richard Fitzpatrick.

of the inverse damping coefficient α^{-1} at fixed values of the parameters $A = 1.5$, $\Omega = 2/3$ and initial conditions $\phi_{t=0} = (d\phi/dt)_{t=0} = 0$. As seen from this figure, at $\alpha^{-1} = 1.245$, the monotonic increase of $d\phi/dt$ reaches a maximum, and then started to decrease. The analysis of the $\phi - d\phi/dt$ phase curve shows [56] that the elliptic trajectories symmetric with respect to the origin are replaced at $\alpha^{-1} = 1.245$ by the asymmetric trajectories, indicating that the pendulum spends more time in the region $\phi < 0$ than in the region $\phi > 0$. Note that this spatial symmetry-breaking solution occurs for the fully symmetric nonlinear pendulum equation. The explanation for this strange behavior is as follows: at $\alpha^{-1} = 1.245$, the nonlinear system with fixed control parameters starts to be sensitive to the change of initial conditions. In Fig. 2.7, the lower curve shows the Poincare section of the trajectory started at $\phi_{t=0} = 0$, $(d\phi/dt)_{t=0} = -3$. For this curve, the phase trajectory is also non-symmetric, being the mirror image of the trajectory related to the upper curve of Fig. 2.7. Therefore, the two curves of Fig. 2.7 correspond to the left and right favoring attractors, respectively, and this symmetry-breaking occurs (in this case) at $\alpha^{-1} = 1.245$.

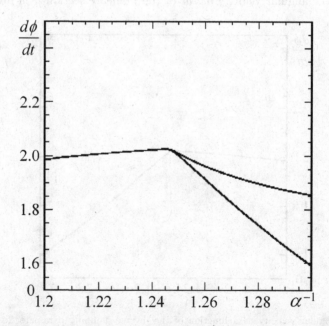

Fig. 2.7   Same as 2.6 with different initial conditions $\phi_0 = 0$, $(d\phi/dt)_0 = -3$. Reprinted with permission from [56]. Copyright (2006) Richard Fitzpatrick.

In Fig. 2.8, we show the two types of regions in $\phi_{t=0} - (d\phi/dt)_{t=0}$ phase space that lead, at $t \to \infty$, to the different attractors shown in Fig. 2.7. The boundary between these two regions is called the separatrix.

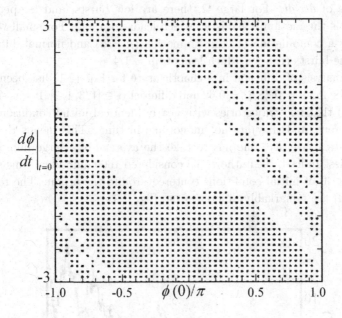

Fig. 2.8 The basin of attractors for a periodically driven damped pendulum. The trajectories which start in the white and dotted regions lead to left-shifted and right-shifted (see text) attractors, respectively. Reprinted with permission from [56]. Copyright (2006) Richard Fitzpatrick.

2.1.6 *Diffusion in a chaotic pendulum*

The characteristic property of deterministic chaos — exponential dependence on initial conditions — leads to an ensemble of trajectories that start out with slightly different initial conditions. These trajectories are equivalent to random walks that exhibit diffusion. Such diffusion can be normal, with a mean-squared amplitude depending linearly on time, or anomalous, with $\langle \phi \rangle^2 \approx t^\mu$ where $\mu \neq 1$. Consider first the Hamiltonian system ($\alpha = 0$ in Eq. (2.1)) [57]. One of the two remaining parameters was kept constant, $A = 1.2$, and the value of μ was calculated for $\Omega = 0.1$, 0.8, 1.4, and 1.5. The initial conditions for 960 trajectories were chosen from a square grid $\phi_{t=0} \in [-0.1, 0.1]$ and $(d\phi/dt)_{t=0} \in [-0.1, 0.1]$. For small Ω ($\Omega \leq 0.8$),

$\mu = 1.48$, whereas for large Ω ($\Omega = 1.4$ and $\Omega = 1.5$), $\mu = 2$ [57]. The chaos considered is of the intermittent type, which means that at this transition, $d\phi/dt$ becomes unstable and periodic oscillations are randomly interrupted by bursts of $d\phi/dt$. For large Ω, there are few bursts, and for periodic oscillations the motion is ballistic, which leads to $\mu = 2$. For small Ω, the motion is a combination of ballistic motion ($\mu = 2$) and normal diffusion during the bursts ($\mu = 1$), which leads to $\mu = 1.4$.

The analysis of diffusion in a chaotic state for Eq. (2.1) has been performed [58] for $A = 0.78, \Omega = 0.62$ and different $\alpha \in [1/3, 1/7]$. It was shown (Fig. 2.9) that two trajectories with nearly identical initial conditions are very different, and they are not monotonic in time. This means that for the statistical average, one has to take the average of a large number of trajectories. Indeed, the authors [58] considered 1600 individual time series with 40×40 of initial conditions centered around the origin. The results (Fig. 2.10) are remarkably different from Fig. 2.9.

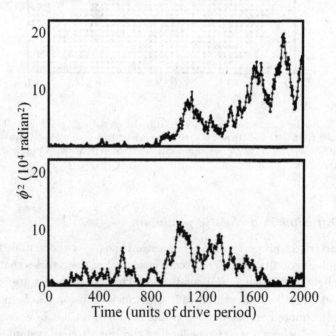

Fig. 2.9 Squared pendulum angle as a function of time for a periodically driven damped pendulum with $A = 0.78$, $\Omega = 0.62$ and $\alpha = 0.25$. Reprinted with permission from [58]. Copyright (1996) by the American Physical Society.

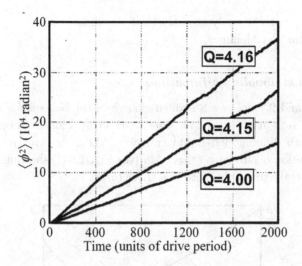

Fig. 2.10 Ensemble average $\langle \phi(t)^2 \rangle$ for three different values of the inverse damping parameter, which shows deterministic diffusion. Reprinted with permission from [58]. Copyright (1996) by the American Physical Society.

Detailed analysis shows that there are many periodic windows embedded within the prevalent chaos. Therefore, the diffusion coefficient is not too sensitive to changes of the parameter α, but it can be considerably enhanced very close to the edge of a periodic window. The results depend on the type of periodic solution. If the latter, which started at some α_{cr}, describes the running solution, corresponding to $\mu = 2$, then the chaotic diffusion coefficient approaches this value proportional to $|\alpha_{cr} - \alpha|^{-1/2}$. If the periodic solutions are bounded in space, which corresponds to zero diffusion coefficient, the chaotic diffusion coefficient approaches this value proportional to $|\alpha_{cr} - \alpha|^{1/2}$. Special attention has been given to the analysis of intermittent chaos which appears at $\alpha^{-1} = 5.78$, where the system randomly switches back and forth between two nonrotating periodic states. The (interrupted) diffusion can occur only for chaotic bursts, which gives $\mu = 1$.

2.2 Analytic methods

The pendulum equations do not have analytic solutions in the chaotic regions, and we have to content ourselves with numerical solutions. There are, however, two types of analytic analysis of the chaotic behavior discussed by Feigenbaum and Melnikov.

2.2.1   *Period-doubling bifurcations*

In Figs. 2.6 and 2.7, we saw a sudden change in the behavior of $d\phi/dt$ as a function of α^{-1} at $\alpha^{-1} = 1.245$, showing the transition from one to two attractors with different left-right symmetry. As α^{-1} increases, another sudden transition occurs at $\alpha^{-1} = 1.348$ (see Fig. 2.11), showing that the left-favoring trajectory splits into two trajectories.

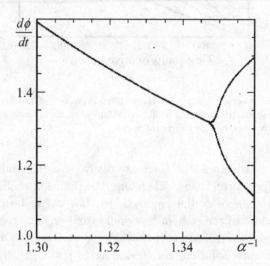

Fig. 2.11   Continuation of Fig. 2.6 for larger values of α^{-1}. Reprinted with permission from [56]. Copyright (2006) Richard Fitzpatrick.

As seen in Fig. 2.12, the same splitting occurs for the right-favoring trajectories at almost the same value of the control parameter α^{-1}. These bifurcations are called the period-doubling, because for $\alpha^{-1} < 1.348$, the period of oscillation is equal to the period of the driving force, and trajectory closes after one revolution, whereas for $\alpha^{-1} > 1.348$, the period is twice the period of the driving force, and the trajectory closes after two revolutions. This period-doubling bifurcation is an example of temporal symmetry breaking which is complimentary to the spatially symmetry breaking at $\alpha^{-1} = 1.245$ considered above. With further increase of the control parameter α^{-1}, this process of frequency doubling is repeated over

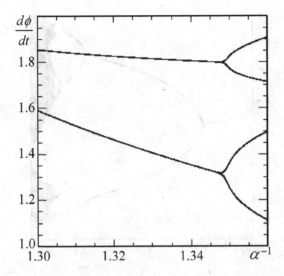

Fig. 2.12   Same as Fig. 2.6, but the data are shown for initial conditions $\phi_0 = (d\phi/dt)_0 = 0$ and $\phi_0 = -\pi/2$, $(d\phi/dt)_0 = -2$ for the lower and upper branches, respectively. Reprinted with permission from [56]. Copyright (2006) Richard Fitzpatrick.

and over (Fig. 2.13): after the first doubling at the first critical value $\alpha_1^{-1} = 1.348$, the second doubling bifurcation (from period 2 to period 4 of the driving force) occurs at $\alpha_2^{-1} = 1.370$, the third (from period 4 to period 8) at $\alpha_3^{-1} = 1.375$, etc. As seen from Fig. 2.13, as n increases, the distance between two succeeding bifurcation $\alpha_n^{-1} - \alpha_{n-1}^{-1}$ becomes smaller and smaller, asymptotically approaching the accumulation point α_∞^{-1}. The latter corresponds to the infinite series of period-doubling bifurcations, which means that the motion never repeats itself, and is thereby chaotic.

The qualitative picture described above was supplemented by a quantitative theory in a remarkable work [59]. It was found that there is an universal (independent on n) ratio of two succeeding values of the parameters defining the period-doublings,

$$\frac{\alpha_{n-1}^{-1} - \alpha_{n-2}^{-1}}{\alpha_n^{-1} - \alpha_{n-1}^{-1}} \equiv \Phi = 4.6692... \tag{2.9}$$

The numerical solution of Eq. (2.1) for different values of the amplitude A of an external field has been performed [60] for $\alpha = 0.5$ and $\Omega = 2/3$. The first period-doubling occurs at $A_1 = 1.073$ and the second occurs at $A_2 = 1.081$. For $A = 1.5$, the motion becomes chaotic. The fact that the de-

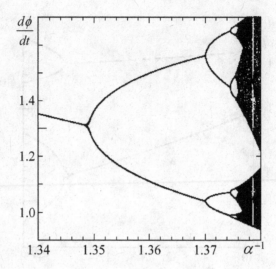

Fig. 2.13 Continuation of Figs. 2.6 and 2.12 showing the full set of period-doubling bifurcations. Reprinted with permission from [56]. Copyright (2006) Richard Fitzpatrick.

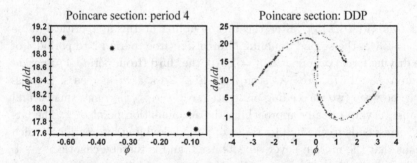

Fig. 2.14 Poincare section for a periodically driven damped pendulum with $\Omega = 2/3$, $\alpha = 0.5$ and initial conditions $\phi_0 = 0$, $(d\phi/dt)_0 = -3$. The left graph shows the regular trajectory of period 4 with $A = 1.081$, whereas the right graph with $A = 1.5$ describes the chaotic motion. Reprinted with permission from [60]. Copyright Tankut Can and Wynton Moore.

terministic chaos appears in deterministic equations means that the initial conditions fully describe the behavior. However, the chaotic solutions are nonperiodic, i.e., they never exactly repeat themselves. For periodic solutions, the Poincare sections contain a discrete number of points (depending on the period of given solution), whereas there are an infinite number of points for a chaotic solution. The latter can be seen by comparing of the Poincare sections in Fig. 2.14.

The left graph, which corresponds to $A_2 = 1.081$, has four points in the phase space displaying graphically the periodic motion with period four, whereas the right graph depicts non-periodic motion for which the points are not expected to repeat.

A most important fact is that the Feigenbaum number Φ describes not only the pendulum, but also the period-doubling transition to chaos in many other systems. Therefore, it is a general property of nonlinear dynamic systems. Feigenbaum began his search for universal numbers by examining the period-doubling cascade in the quadratic map of the form $x_{n+1} = \lambda x_n (1 - x_n)$. With the help of a pocket calculator, he found the values of λ at which the first few period-doubling bifurcations occur (Table 1).

Table 1. Period-doubling bifurcation values for the quadratic map.

$\lambda_1 = 3.0$	$\lambda_4 = 3.5644$
$\lambda_2 = 3.4495$	$\lambda_5 = 3.5687$
$\lambda_3 = 3.5441$	$\lambda_6 = 3.5697$

Based on his calculations, Feigenbaum found [61] that

$$\frac{\lambda_{n-1} - \lambda_{n-2}}{\lambda_n - \lambda_{n-1}} = \Phi; \qquad \lambda_\infty - \lambda_n = \frac{a}{\Phi^n} \qquad (2.10)$$

where the three universal constants in these equations are $\Phi = 4.6692$, $\lambda_\infty = 3.5698$ and $a = 2.637$.

In the following sections, we present the results of numerical calculations of period-doubling bifurcations for different types of pendulum equations.

2.2.2 Melnikov method

A comprehensive presentation of the Melnikov method has been given [62]. Here we apply this method to Eq. (2.1) of the damped driven pendulum. To use the Melnikov method, one must rewrite Eq. (2.1) as two first-order differential equations,

$$\frac{d\phi}{dt} = y; \qquad \frac{dy}{dt} = -\sin\phi - \varepsilon\alpha y - \varepsilon A \sin(\Omega t) \qquad (2.11)$$

For $\varepsilon = 1$, these equations reduce to Eq. (2.1), and for $\varepsilon = 0$, they represent the unperturbed system,

$$\frac{d\phi}{dt} = y; \qquad \frac{dy}{dt} = -\sin\phi, \qquad (2.12)$$

The latter equations have a saddle-type equilibrium for $\phi = \pm\pi$, and the center-type equilibrium position for $\phi = 0$. The trajectory that begins and ends in the saddle point surrounding the center is called the homoclinic trajectory. If a pendulum has the potential energy of the upward position, then the pendulum will be moving on the homoclinic trajectory. The trajectory that connects two bifurcation points is called the heteroclinic trajectory. Dynamic systems are characterized by attractors, which attract the trajectories from a set of initial conditions (basin of attraction). If system has two (or more) attractors, the boundary between different attractors is called the separatrix. For two attractors, the separatrix is a simple curve. The two basins of attraction (black and white regions) and the separatrix are shown in Fig. 2.15 [63].

This curve has a fractal structure which means that analyzing the basin boundary of successively smaller and smaller region of the phase space yields the same structure (self-similarity). For a non-dissipative system, the separatrix may split as result of a perturbation.

In Fig. 2.16, we show [64] the separatrix for an unperturbed system, described by Eqs. (2.12), and three possible forms of the separatrix for a perturbed system (2.11). Only the last case (with many intersections) describes chaotic behavior. The Melnikov function [65] defines the distance between the two parts of a split separatrix, the vanishing of which describes the onset of chaos. In order to use the Melnikov theory, one has to find the heteroclinic trajectories $\phi_h(t)$ for the unperturbed problem, and then substitute this function in the Melnikov integral,

$$M(t_0) = \int_{-\infty}^{\infty} \phi_h(t - t_0) F(t - t_0) dt \qquad (2.13)$$

where εF is the perturbation described by Eq. (2.11). In our case,

$$\phi_h(t) = \pm 2\arctan(\sinh t); \qquad (2.14)$$

$$M(t_0) = \int_{-\infty}^{\infty} y_h(t - t_0)[A\sin(\Omega t) - \alpha y_h(t - t_0)] dt \qquad (2.15)$$

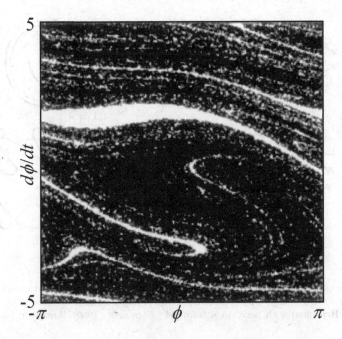

Fig. 2.15   Typical basins of two attractors (shown in black and white) with the boundary (separatrix) between them. Reprinted from [63], Copyright (1987), with permission from Elsevier.

which gives [66]

$$M\left(t_0\right) = -8\alpha \pm 2\pi A\mathrm{sech}\left(\pi\Omega/2\right)\sin\left(\Omega t_0\right) \tag{2.16}$$

The vanishing of the Melnikov function (2.16) defines the occurrence of homoclinic points, which are "precursors of chaos". The actual chaos threshold is located a little bit over the Melnikov criterion. The Melnikov method is a perturbation approach, which, strictly speaking, is correct, only in the limit $\varepsilon \ll 1$. With these constraints, Eq. (2.16) gives the Melnikov criterion for the onset of chaos,

$$A \geq \frac{4\alpha}{\pi}\cosh\left(\pi\Omega/2\right) \tag{2.17}$$

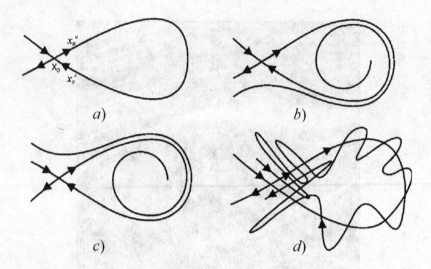

Fig. 2.16 Possible forms of the Melnikov function $M(t_0)$: a) $M(t_0) = 0$ (unperturbed system), b) $M(t_0) < 0$, c) $M(t_0) > 0$. d) $M(t_0)$ changes its sign at some t_0 (chaotic behavior). Reprinted with permission from [64], Copyright (1980), American Institute of Physics.

2.3 Parametric periodic force

Thus far, we have considered a pendulum whose suspension point is at rest. The motion of the suspension point gives rise to a multiplicative force in the dynamic equation, unlike the external force which enters the pendulum equation of motion additively. When the suspension point performs vertical oscillations $u(t) = a \sin \omega t$, the system becomes non-inertial, and it is convenient to use a non-inertial frame of reference fixed to the moving axis. Newton's law must then be modified by the addition of the force of inertia $-m d^2 u/dt^2$. This means that for a pendulum with a vertically oscillating suspension point, one has to replace the gravity g by $g - d^2 u/dx^2$, which leads to the following dimensionless equation

$$\frac{d^2\phi}{dt^2} + \alpha \frac{d\phi}{dt} + [1 + A\sin(\Omega t)]\sin\phi = 0 \tag{2.18}$$

When the suspension point executes periodic horizontal oscillations, similar arguments lead to the following equation

$$\frac{d^2\phi}{dt^2} + \alpha \frac{d\phi}{dt} + \sin\phi + A\sin(\Omega t)\cos\phi = 0 \tag{2.19}$$

We will first consider the vertical and horizontal oscillations of the suspension point separately, and then the general case when the suspension point moves harmonically in both the horizontal and vertical directions, leading to the following equation of motion

$$\frac{d^2\phi}{dt^2} + \alpha\frac{d\phi}{dt} + A_1 \sin(\Omega_1 t)\cos\phi + [1 + A_2 \sin(\Omega_2 t)]\sin\phi = 0 \qquad (2.20)$$

2.3.1 *Pendulum with vertically oscillating suspension point*

Equation (2.18) cannot be solved analytically. A numerical analysis of Eq. (2.18) was performed in 1981 by McLaughlin [67] for $\Omega = 2$ and for three values of α (0.2, 0.05 and 0). In order to obtain motion with a finite amplitude, the control parameter A has to satisfy the condition $A \geq \alpha/2$ [68]. For $\alpha = 0.2$, simple vibration of period 2π is the only solution for $1.426 > A > 0.4$. At $A = 1.426$, two rotations of period π appear. Then, at $A = 1.582$, these rotations go through a period-doubling bifurcation, until finally, a pair of strange attractors appear. Up to $A = 2.088$, the sign of $d\phi/dt$ is negative (clockwise rotation), but for larger A, the angular velocity occasionally switches sign. The strange attractors exist up to $A = 3.16$. Beyond this point, there are two stable solutions of period π. Thus, increasing the control parameter A causes the system to go through the transformation order-chaos-order. Qualitatively similar behavior takes place for the damping constant $\alpha = 0.05$. However, no strange attractors exists for the Hamiltonian case ($\alpha = 0$), and the double-frequency bifurcation leads to the destruction of stable zones.

The numerical solution of Eq. (2.18) has been obtained for different sets of parameters [69]. For $\alpha = 0.15$ and $\Omega = 1.56$, the results are similar to those obtained in [67], with a fully chaotic regime for $A > 0.458$. The behavior of a system near the latter point depends on the initial conditions. The onset of chaos has been checked by an analysis of the power spectrum and the Lyapunov exponents [69]. These results has been confirmed by experiment [70].

2.3.2 *Transition to chaos*

Depending on the values of parameters α, A and Ω in Eq. (2.18) and the initial conditions, the pendulum can undergo different types of dynamic behavior, such as oscillations, rotations, which, upon increasing the parameters, lead to oscillating and rotating chaotic solutions. A comprehensive analysis has been performed [71] for $\alpha = 0.1$ and $\Omega = 2$. Oscillating chaos appears at $A = 1.3426$, whereas rotating chaos settles down at $A = 1.81$. In addition to these two types of chaotic motions, there is a third type of chaos, termed "tumbling chaos." The latter is an irregular combination of rotation and oscillation, in which the pendulum completes an apparently random number of clockwise (or anti-clockwise) rotations before changing direction, while performing a number of oscillations about the hanging position. The tumbling attractor appears at $A = 2$.

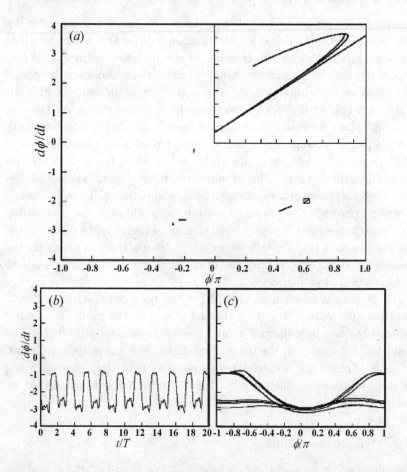

Fig. 2.17   Rotating chaos for a parametrically driven damped pendulum with $A = 1.81$, $\Omega = 2$, and $\alpha = 0.1$. The three graphs show the Poincare section, the time dependence of the angular velocity and the phase diagram. Reprinted from [71]. Copyright (1996), with permission from Elsevier.

The results of the analysis of rotating chaos for $A = 1.81$ and $\Omega = 2$ are shown in Fig. 2.17, which exhibits the Poincare section, the time dependence of the angular velocity and the phase diagram [71]. It turns out that the zones of oscillating and rotating chaos occupy only a very small part of parametric space, whereas tumbling chaos occurs over a broad band of parameters. The fact that this region is comparatively large indicates that the response is robust with respect to small changes in the system parameters, whereas oscillating and rotating chaotic motions are less structurally stable.

The three different types of chaos described above can be characterized by the rotation number [72],

$$\rho \equiv \frac{\langle d\phi/dt \rangle}{\Omega} = \lim_{t \to 0} \frac{\phi(t) - \phi(0)}{\Omega t} \qquad (2.21)$$

The rotation number ρ vanishes identically for periodic oscillations and takes non-zero values for periodic rotations. For tumbling chaos, the rotation number is not well defined, being subject to fluctuations, but one may define the rotation interval over which ρ fluctuates. Viewing the rotating number as an order parameter, the results of the numerical solution of Eq. (2.18) can be used to construct the $\rho - A$ phase diagram, which was further clarified in combination with the analysis of bifurcation diagrams.

The period-doubling transition to chaos in Eq. (2.18) for $\alpha = 0.2$ and $\Omega = 2$ has been discussed [67]. The vibration with the period of a driving field starts at $A > 0.4$, and this vibration is the only solution up to $A_1 = 1.426 \pm 0.024$ [67]. At A_1, two distinct (clockwise and counterclockwise) rotational motions appear. Each goes through an additional period-doubling at $A_2 = 1.5850 \pm 0.0050$, etc. Finally [67],

$$\frac{A_2 - A_1}{A_3 - A_2} = 5.35 \pm 0.30; \quad \frac{A_3 - A_2}{A_4 - A_3} = 3.25 \pm 0.25; \quad \frac{A_4 - A_3}{A_5 - A_4} = 5.6 \pm 1.0$$
$$(2.22)$$

It is possible that this sequence is described by the Feigenbaum constant $\Phi = 4.669$. The end result of cascade of period-doubling bifurcations is a pair of strange attractors.

2.3.3 Melnikov method.

The Melnikov method is an approximate procedure which allows one to find analytically the region of parameters where chaos occurs. The governing equation (2.18) can be rewritten

$$\frac{d\phi}{dt} = y; \qquad \frac{dy}{dt} = -\varepsilon\alpha y - [1 - \varepsilon A \sin{(\Omega t)}]\sin\phi \qquad (2.23)$$

where the small parameter $\varepsilon \ll 1$ permits the use of perturbation theory. Some questions that should be answered for fixed values of parameters Ω and α are the following:

1. For which values of driving amplitude A does chaos occur?

2. How small do the damping and the driving force have to be to justify the perturbation procedure? Equivalently, how small does the parameter ε have to be?

3. How important is the division of the equation of motion into the unperturbed problem ($\varepsilon = 0$) and perturbation?

Here, we will answer the first two questions, leaving the third question to Section 2.5.

For homoclinic orbits, the Melnikov function is

$$M(t_0) = 2\pi A\Omega^2 \text{cosech}\,(\pi\Omega/2)\sin{(\Omega t_0)} - 8\alpha \qquad (2.24)$$

which gives the following condition on the driving amplitude for the occurrence of chaos,

$$\frac{A}{\alpha} \geq \frac{4}{\pi\Omega^2}\sinh\left(\frac{\pi\Omega}{2}\right), \qquad (2.25)$$

For $\Omega = 1.56$, $(A/\alpha)_{\min} = 3.0102$. This result can be compared with the numerical results in Table 2 [73]. It is seen that for these values of the "small" parameter ε, the Melnikov criterion gives quite reasonable results.

Table 2. Numerical parameters for the bifurcation.

ε	A/α
0.15	3.0097 ± 0.003
0.3	3.0083 ± 0.003
0.5	3.0081 ± 0.001

2.3.4 *Parametric periodic non-harmonic force*

Thus far, we have restricted our discussion to a pendulum subject to harmonic vertical oscillations of the suspension point. Harmonic functions are the simplest solution of the linear harmonic oscillator, and they are usually

used as an external force for the linear oscillator leading to different resonant phenomena. For the nonlinear dynamics considered here, the natural
choice of the driving force is the eigenfunction of the appropriate nonlinear
problem. As seen in Section 1.2, for the pendulum, those are the Jacobi elliptic functions. The elliptic functions $f(t, m)$ belong to a class of periodic
functions $f(t, m) = f(t + T, m)$ having parameter m. The most common
elliptic functions are sine and cosine amplitude Jacobi functions $sn(\Omega t, m)$
and $cn(\Omega t, m)$ with $0 \leq m \leq 1$. For the limiting values $m = 0$ and $m = 1$,
$sn(\Omega t, m)$ and $cn(\Omega t, m)$ reduce to the trigonometric functions, $\sin(\Omega t)$ and
$\cos(\Omega t)$, and to the hyperbolic functions, $sech(\Omega t)$ and $\tanh(\Omega t)$, respectively.

The generalization of Eq. (2.18) to the case for which the suspension
point executes a vertical periodic displacement, $cn(\Omega t, m)$ of amplitude A
and frequency Ω satisfies the following equation.

$$\frac{d^2\phi}{dt^2} + \alpha\frac{d\phi}{dt} + [1 + Acn(\Omega t, m)]\sin\phi = 0 \qquad (2.26)$$

The numerical analysis of Eq. (2.26) has been performed [74] for $\alpha = 0.1$
and $\Omega = 1.5$. The results of the calculation of the chaotic parametric set
are shown in Fig. 2.18.

The Lyapunov exponents have been calculated for each point on this
$m - A$ plane. The points with a positive Lyapunov exponent (chaotic
regions) are black and those with a negative exponent are white. One
may consider the horizontal lines as a kind of bifurcation diagram. As
an example, the line $m = 0$ corresponds to the harmonic function in Eq.
(2.26), which reduces to Eq. (2.18). On the other hand, one can draw a
vertical line in Fig. 2.18, and vary m from zero to one. For example, a
vertical line at $A = -4$ crosses the black region only in a small range of
values of m close to $m = 1$, i.e., the majority of trajectories for this value
of A are non-chaotic.

The Melnikov method has been used [75] for the analysis of Eq. (2.26)
which can be rewritten as

$$\frac{d\phi}{dt} = y; \qquad \frac{dy}{dt} = -\sin\phi - \varepsilon\alpha y - \varepsilon Acn(\Omega t) \qquad (2.27)$$

The Melnikov function calculated for the homoclinic orbit is

$$M(t_0) = 2\int_{-\infty}^{\infty} sech[h(t - t_0)][-Acn(\Omega t, m)\sin(t - t_0) - \alpha sech(t - t_0)]\,dt \qquad (2.28)$$

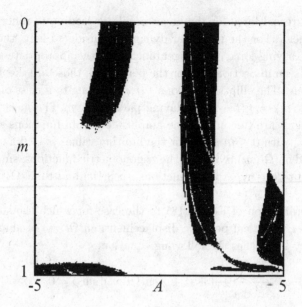

Fig. 2.18 Chaotic diagram for a pendulum with a periodic non-harmonic force in the $m - A$ plane, where A is the amplitude and m is the characteristic parameter of the Jacobi elliptic function. The white points correspond to a negative Lyapunov exponent, whereas the black points correspond to a positive Lyapunov exponent (chaotic behavior). Reprinted with permission from [74]. Copyright (1998) by the American Physical Society.

After calculating the appropriate integrals, the Fourier expansion of $cn(\Omega t, m)$ yields, the Melnikov criterion for the onset of chaos [75],

$$\frac{A}{\alpha} \leq J^{-1}(\Omega, m) \qquad (2.29)$$

where

$$J_1(\Omega, m) = \frac{\pi^4 \Omega^2}{16 m K^3} \sum_{j=0}^{\infty} (2j+1)^2 \, \text{sech} \left[\frac{(2j+1)\pi K_1}{2K} \right]$$

$$\times \text{csch} \left[\frac{(2j+1)\pi^2 \Omega}{4K} \right] \qquad (2.30)$$

and $K(m)$ is is the complete elliptic integral of the first kind and $K_1 = K\left(\sqrt{1-m^2}\right)$

For $m = 0$, the criterion (2.29) reduces to (2.25), which was obtained for a pendulum with vertically harmonic oscillations of the suspension point.

2.3.5 Downward and upward equilibrium configurations

The prediction of stability for the upward position of a pendulum subject to vertical oscillations of the suspension point, was obtained more than a century ago [76]. We shall explain physically [77] why rapid vertical oscillations of the suspension point stabilize the inverted (upward) position of a pendulum. In Eq. (2.18), the torque of the gravitational force tends to tip the pendulum downward while the torque of the inertial force (averaged over the period Ω of rapid oscillations) tends to return the pendulum to the inverted position. If the latter is large enough, the inverted position of a pendulum will be stable. The quantitative criterion for such stabilization can be obtained in the following way [77]. One starts with the conjecture that the angle ϕ is the superposition of two components

$$\phi = \phi_{slow} + \phi_{fast}, \tag{2.31}$$

implying

$$\sin \phi \approx \sin \phi_{slow} + \phi_{fast} \cos \phi_{slow} \tag{2.32}$$

where the "slow" angle, ϕ_{slow}, has a small variation during a period of constrained oscillations, whereas the "fast" angle, ϕ_{fast}, is small with zero mean value, $\langle \phi_{fast} \rangle = 0$. The angle ϕ_{fast} oscillates with high frequency Ω and has an amplitude proportional to the sine of the momentary value of $\phi = \phi_{slow}$,

$$\phi_{fast} = -A \sin \phi_{slow} \sin (\Omega t) \tag{2.33}$$

According to this equation, the average value of the gravitational torque $\langle \sin \phi \rangle = \langle \sin \phi_{slow} \rangle$, which is the same as for a pendulum with a fixed suspension point, whereas the average value of the inertial torque

$$\langle A \sin \phi \sin (\Omega t) \rangle = \langle A \sin (\phi_{slow} + \phi_{fast}) \sin (\Omega t) \rangle \approx$$
$$\approx \langle A \sin \phi_{slow} + A \phi_{fast} \cos \phi_{slow} \sin (\Omega t) \rangle \tag{2.34}$$

contains an additional term which is equal to $\frac{1}{2} A \sin \phi_{slow} \cos \phi_{slow}$, where Eq. (2.32) and $\overline{\sin^2 (\Omega t)} = \frac{1}{2}$ have been used. Comparing the latter term with the gravitational torque shows that the inertial torque can exceed the gravitational torque and cause the pendulum to tip up when the following condition is fulfilled,

$$A^2 \Omega^2 > 1 \tag{2.35}$$

The latter formula can be rewritten for the dimensional amplitude and frequency,

$$\frac{1}{2}\frac{A^2}{l^2}\frac{\Omega^2}{\omega_0^2} > 1 \tag{2.36}$$

Adding damping in Eq. (2.18) causes the stability of the inverted state to decrease while the stability of the downward position is enhanced [78].

The comprehensive numerical solution of Eq. (2.18), written in a slightly different form,

$$\frac{d^2\phi}{dt^2} + \alpha\frac{d\phi}{dt} + \left[\omega_0^2 + A\sin(\Omega t)\right]\sin\phi = 0$$

has been obtained by Bartuccelli et al. [79]. For parameters $\alpha = 0.03$, $\omega_0^2 = 0.5$, and $\Omega = 1$, a period-doubling transition to chaos occurs when the parameter A is increased above 0.55, corresponding to the downward stable point. For $A_{cr} = 0.640$, the motion is fully chaotic. On the other hand, for $\alpha = 0.08$, $\omega_0^2 = -0.1$, and $\Omega = 1$, the upward position is stable for $A < A_{cr} = 0.623$, when chaos occurs. It was also found [79] that different attractors fill the entire phase space. In other words, every initial boundary condition will eventually end up in one of the attractors.

2.3.6   *Boundary between locked and running solutions*

Of great importance for many applications are the values of the parameters that determine the boundaries between locked and running solutions ("escape parameter region"). This problem is analogous to the escape of a particle from a potential well. To solve this problem for the parametrically excited damped pendulum, one uses the harmonic balance method [80]. This yields the following dimensionless equation defining the transition from the locked to the running solution in the $\Omega - A$ plane

$$\left(\frac{\Omega^2}{4} - 1\right)^2 + \frac{\alpha^2\Omega^4}{4} - \frac{A^2}{4} = 0 \tag{2.37}$$

An unconventional approach has recently been taken [81] to obtain the analytical solution of the problem of the locked-running transition by transforming the differential equation (2.18) into an integral equation. Multiplying Eq. (2.18) by $d\phi/dt$ yields the law of the conservation of energy,

$$\frac{d}{dt}\left[\frac{1}{2}\left(\frac{d\phi}{dt}\right)^2 - \cos\phi\right] = -\alpha\left(\frac{d\phi}{dt}\right)^2 - A\frac{d\phi}{dt}\cos\left(\Omega t\right)\sin\phi \qquad (2.38)$$

Integrating this equation leads to

$$\left(\frac{d\phi}{dt}\right)^2 - 2\cos\phi - \left(\frac{d\phi}{dt}\right)_0^2 - 2\cos\phi_0 = \qquad (2.39)$$

$$-2\alpha\int_{t_0}^{t}\frac{d\phi\left(r\right)}{dr}\frac{d\phi}{dr}dr - 2A\int_{t_0}^{t}\cos\left(\Omega r\right)\sin\left(\left[\phi\right]r\right)\frac{d\phi}{dr}dr$$

where ϕ_0 and $(d\phi/dt)_0$ are the initial conditions at time t_0. For rotations, $d\phi/dt \neq 0$, and one can find the inverse function $t = t\left(\phi\right)$. For the function $d\phi/dt = \psi\left[\phi\left(t\right)\right]$, the inverse function is

$$t = t_0 + \int_{t_0}^{t}\frac{ds}{\psi\left(s\right)} \qquad (2.40)$$

One rewrites Eq. (2.39) as

$$\psi\left(\phi\right)^2 - 2\cos\phi - \left(\frac{d\phi}{dt}\right)_0^2 - 2\cos\phi_0 =$$

$$-2\alpha\int_{\phi_0}^{\phi}\psi\left(s\right)ds - 2A\int_{\phi_0}^{\phi}\cos\left(\Omega t_0 + \Omega\int_{\phi_0}^{z}ds/\psi\left(s\right)\right)\sin z\,dz \qquad (2.41)$$

For rotations with period T,

$$\phi\left(T\right) = \phi\left(t_0 + T\right) = \phi_0 + 2\pi; \quad \frac{d\phi}{dt}\left(T\right) = \frac{d\phi}{dt}\left(t_0 + T\right) = \left(\frac{d\phi}{dt}\right)_0, \qquad (2.42)$$

Using Eq. (2.42), one can rewrite Eq. (2.39) for $\phi = \phi_0 + 2\pi$,

$$\alpha B + A\cos\left(\Omega t_0\right)C - A\sin\left(\Omega t_0\right)D = 0, \qquad (2.43)$$

where

$$B = \int_{\phi_0}^{\phi_0 + 2\pi}\psi\left(s\right)ds; \quad C = \int_{\phi_0}^{\phi_0 + 2\pi}\cos\left(\Omega\int_{\phi_0}^{z}ds/\psi\left(s\right)\right)\sin z\,dz$$

$$D = \int_{\phi_0}^{\phi_0 + 2\pi}\sin\left(\Omega\int_{\phi_0}^{z}ds/\psi\left(s\right)\right)\sin z\,dz \qquad (2.44)$$

Defining the angle β by

$$\sin \beta = \frac{C}{\sqrt{C^2 + D^2}}; \qquad \cos \beta = \frac{D}{\sqrt{C^2 + D^2}}, \qquad (2.45)$$

Eq. (2.43) becomes

$$\sin(\Omega t_0 - \beta) = \frac{\alpha}{A} \frac{B}{\sqrt{C^2 + D^2}} \qquad (2.46)$$

Equation (2.46) implies that a periodic solution exists if and only if the amplitude A of the external force satisfies

$$A \geq A_{cr} = \frac{\alpha B}{\sqrt{C^2 + D^2}} \qquad (2.47)$$

Equation (2.47) was obtained by an exact analytical calculation for the function $\psi(\phi)$ which defines the trajectory $d\phi/dt = \psi(\phi)$ in phase space. This approach leads to a convenient form of perturbation theory [81].

2.3.7 Pendulum with horizontally oscillating suspension point

Equation (2.19) is symmetric with respect to the following transformations

$$\phi \longrightarrow -\phi; \qquad \frac{d\phi}{d\tau} \to -\frac{d\phi}{d\tau}; \qquad \tau \to \tau + \frac{\pi}{2} \qquad (2.48)$$

The trajectory in phase space $(\phi,\ d\phi/dt,\ t)$ that is invariant under these symmetry transformations is called a symmetric trajectory. Otherwise, it is called an asymmetric trajectory. Detailed numerical analysis of Eq. (2.19) has been performed [82] for different values of the amplitude A of an external field in the interval $A \in (0, 15)$, for $\alpha = 0.1$ and $\Omega = 0.8$. Five cascades of double-frequency bifurcations were found for specific values of A which satisfy the Feigenbaum theory [59]. The numerical solution of Eq. (2.19) shows that upon increasing the control parameter, the system executes a series of the period-doubling bifurcations before the transition to chaos. The control parameter F used in [82] is related to our control parameter A by $A = 0.64\,F$. The first five bifurcations for the subsequent period-doublings occur at $F_1 \approx 1.430786$, $F_2 \approx 1.441438$, $F_3 \approx 1.443138$, $F_4 \approx 1.443480$, $F_5 \approx 1.443553$, which gives the following results for the Feigenbaum number Φ,

$$\frac{F_2 - F_1}{F_3 - F_2} = 6.2681; \qquad \frac{F_3 - F_2}{F_4 - F_3} = 4.9696; \qquad \frac{F_4 - F_3}{F_5 - F_4} = 4.6686 \qquad (2.49)$$

approaching the theoretical prediction $\Phi = 4.6692$.

Equations (2.49) describe only the initial part of this diagram near the first cascade of bifurcation. The entire order-to-chaos transition is much more complicated. It shows the appearance of additional cascades of the period that are three times as large as the period of the external force with its subharmonics that appear near $F = 10.27$ and 10.75. In addition, there are several chaotic bands for $1.44 < F < 1.62$, a small band near $F = 7.01$, a band for $10.46 < F < 10.75$, etc. The periodic and quasi-periodic solutions lie between these bands of chaotic solutions.

At the limiting value $F = 1.4436$, the behavior of the system becomes chaotic. The appearance of frequency-doubling and the transition to chaos can be clearly seen in each cascade. For the first cascade, the transitions of periodic orbit $\to$ frequency doubling $\to$ chaos can be seen from the $(\phi, d\phi/dt)$ phase plane shown in Fig. 2.19.

Additional information has been obtained [82] by studying the Lyapunov exponents, the power spectrum and the evolution of strange attractors upon changing the control parameter A. The solutions of Eq. (2.19) have been studied numerically for different frequencies of an external field at constant amplitude [83]. As the frequency is decreased, at sufficiently large amplitude, the system progresses from symmetric trajectories to a symmetry-breaking period-doubling sequence of stable periodic oscillations, and finally to chaos.

The analysis of Eq. (2.19) has been performed [84] for the special case of small amplitude and high frequency of the external field. The latter conditions are satisfied if one replaces the frequency Ω of the external field by ω such that $\omega = \Omega/\varepsilon$. The small parameter ε also determines the magnitude of the amplitude, $A = \varepsilon\beta$. One obtains $2\alpha\varepsilon$ for the damping coefficient. In terms of these new parameters and the dimensionless time $\tau = \Omega t$, Eq. (2.19) takes the following form

$$\frac{d^2\phi}{d\tau^2} + 2\alpha\varepsilon\frac{d\phi}{d\tau} - \beta\varepsilon\sin\tau\cos\phi + \varepsilon^2\sin\phi = 0 \qquad (2.50)$$

According to the method of multiple scales [85], one seeks the solution of Eq. (2.50) of the form

$$\phi = \phi_0 + \varepsilon\phi_1 + \varepsilon^2\phi_2 + \dots \qquad (2.51)$$

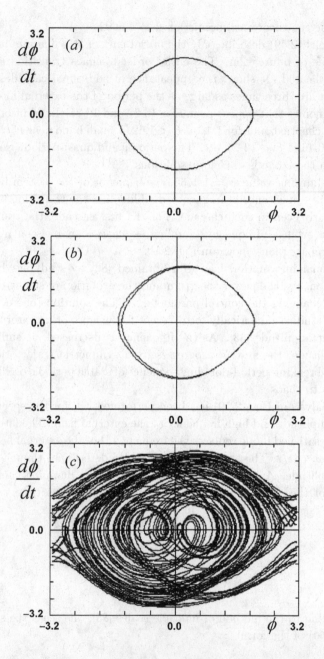

Fig. 2.19 Period-doubling bifurcations for a horizontally driven pendulum: a) Period-1 solution at $F = 1.429$, b) Period-2 solution at $F = 1.4359$, c) Chaotic solution at $F = 1.45$. Reprinted from [82], Copyright (1996), wih permission from Elsevier.

where ϕ_n is a function of $\varepsilon^n \tau_n$. Then,

$$\frac{d}{d\tau} = D_0 + \varepsilon D_1 + \varepsilon^2 D_2 + ...;$$

$$\frac{d^2}{d\tau^2} = D_0^2 + 2\varepsilon D_0 D_1 + 2\varepsilon^2 D_0 D_2 + \varepsilon^2 D_1^2 ... \tag{2.52}$$

where $D_n \equiv d/d\tau_n$ Inserting (2.51) and (2.52) into (2.50) yields three equations for the coefficients of ε^0, ε^1 and ε^2,

$$D_0^2 \phi_0 = 0 \tag{2.53}$$

$$D_0^2 \phi_1 = -2D_0 D_1 \phi_0 + \beta \sin \tau_0 \cos \phi_0 - 2\alpha D_0 \phi_0 \tag{2.54}$$

$$D_0^2 \phi_2 = -2D_0 D_1 \phi_1 - 2D_0 D_2 \phi_0 - D_1^2 (\phi_0 - \sin \phi_0)$$
$$-2\alpha D_0 \phi_1 - 2\alpha D_1 \phi_0 - \beta \phi_1 \sin \tau_0 \cos \phi_0 \tag{2.55}$$

The solution of Eq. (2.53) that does not increase with time is $\phi_0 = Const.$ Using Eqs. (2.54) and (2.55), and eliminating the secular (unbounded) terms yields three solutions for the stationary state, $\phi_0 = 0$, $\phi_0 = \pi$ and $\phi_0 = \cos^{-1}(2/\beta^2)$. Stability analysis [84] shows that the fixed-point solution $\phi_0 = \pi$ is always unstable, whereas for $A/l \leq \sqrt{2}\omega/\Omega$, the only stable point is $\phi = 0$. When the amplitude of the external force β increases beyond $\sqrt{2}\omega l/\Omega$, the zero solution becomes unstable, and the two stable solutions are $\pm \cos^{-1}(2/\beta^2)$. This analytic result, which is supported by the numerical solution of Eq. (2.50), is quite surprising. Indeed, the pendulum with horizontal oscillations of the suspension point does not perform oscillations around the horizontal axis, but rather around an inclined axis!

One can also apply the Melnikov method to analyze the horizontal motion of the suspension point. We rewrite Eq. (2.19)

$$\frac{d\phi}{dt} = y; \qquad \frac{dy}{dt} = -\sin \phi - \varepsilon \alpha \frac{d\phi}{dt} - \varepsilon A \sin(\Omega t) \cos \phi \tag{2.56}$$

The Melnikov function is [86]

$$M(t_0) = -8\alpha - 2\pi A \Omega^2 \, \text{sech} \left(\frac{\pi \Omega}{2}\right) \cos(\Omega t_0), \tag{2.57}$$

with the following amplitude A of the external force which defines the occurring of chaos,

$$\frac{A}{\alpha} \geq \frac{4}{\pi \Omega^2} \cosh \left(\frac{\pi \Omega}{2}\right) \tag{2.58}$$

Analyzing this equation shows [86] that it is more difficult to find a region in which system evolves into a nonchaotic regime when the (linearized) eigenfrequency is raised for a given driving frequency. In other words, for given amplitude of the driving force, suppression of chaos is easier when the pendulum eigenfrequency is lower.

2.3.8   *Pendulum with both vertical and horizontal oscillations of the suspension point*

The Melnikov method can also be applied to the analysis of equation of motion (2.20). In the general case, the trajectory of the suspension point, subject to two orthogonal harmonic oscillations, is quite complicated. For simplicity, consider the case $A_1 = A_2 \equiv A$ and $\Omega_1 = \Omega_2 \equiv \Omega$, for which the motion is circular. Then, the Melnikov function is [86]

$$M(t_0) = -8\alpha - 2\pi A\Omega^2 \left[\operatorname{sech}\left(\frac{\pi\Omega}{2}\right) + \operatorname{csch}\left(\frac{\pi\Omega}{2}\right) \right] \cos(\Omega t_0) \qquad (2.59)$$

The vanishing of the Melnikov function (2.24) defines the "precursors of chaos," and the amplitude A of the external force which defines the occurrence of chaos is defined as

$$\frac{A}{\alpha} \geq \frac{4}{\pi\Omega^2} \left[\operatorname{sech}\left(\frac{\pi\Omega}{2}\right) + \operatorname{csch}\left(\frac{\pi\Omega}{2}\right) \right]^{-1} \qquad (2.60)$$

It is interesting to compare the results of the Melnikov analysis of a pendulum subject to an external force (Eq. (2.25)) with that of a pendulum with a moving suspension point (Eq. (2.60)). In the latter case, chaos occurs at smaller amplitude A [86].

2.4   Parametrically driven pendulum

An external excitation acting on a driven pendulum may be introduced in three different ways. In addition to an external and a parametric periodic force, considered in previous chapters, one can consider chaotic solutions of a parametrically driven pendulum described by the following equation

$$\frac{d^2\phi}{dt^2} + \alpha\left[1 + A\sin(\Omega t)\right]\frac{d\phi}{dt} + \sin\phi = 0 \qquad (2.61)$$

In contrast to Eq. (2.1) that describes a pendulum subject to an external force, Eq. (2.61) admits a stationary solution $\phi = d\phi/dt = 0$. However, this solution is not necessarily stable for arbitrary values of the parameters A and Ω. To test for stability, one considers a small derivation from the origin described by small ϕ and $d\phi/dt$, providing the damping parameter α equal to $(18.63)^{-1}$ [87]-[89]. Then, $\sin\phi \approx \phi$, and the last two terms in Eq. (2.61) are small. Following the Lindsted-Poincare perturbation procedure, one introduces $\varepsilon = \alpha\Omega$ and $\Omega t = \tau$, which converts Eq. (2.61) into

$$\Omega^2 \frac{d^2\phi}{d\tau^2} + \varepsilon\left[1 + A\sin(\tau)\right]\frac{d\phi}{d\tau} + \phi = 0 \qquad (2.62)$$

Expansion in the small parameter ε

$$\Omega = \Omega_0 + \varepsilon\Omega_1 + \varepsilon^2\Omega_2 + \dots$$
$$\phi = \phi_0 + \varepsilon\phi_1 + \varepsilon^2\phi_2 + \dots \qquad (2.63)$$

yields a set of equations for ϕ_0 and ϕ_1 that are related to each other. The solutions of these equations to order ε^2 define the stability boundary for Eq. (2.61) [87],

$$\Omega = 2 + \alpha\left(\frac{A^2}{4} - 1\right)^{1/2} - \alpha^2\left(\frac{A^2}{16} + \frac{1}{4}\right), \quad \Omega > 2$$

$$\Omega = 2 - \alpha\left(\frac{A^2}{4} - 1\right)^{1/2} - \alpha^2\left(\frac{A^2}{16} + \frac{1}{4}\right), \quad \Omega < 2 \qquad (2.64)$$

and $A = 2$ for $\Omega = 2$.

Another way to find the boundary of stability for Eq. (2.62) is to use the Floquet theorem which defines the existence of a periodic solution for ϕ [89]. Assuming the solution of Eq. (2.62) in the form

$$\phi = \phi_0 + \sum_{n=1}^{\infty}(A_n\cos nt + B_n\sin nt) \qquad (2.65)$$

and inserting (2.65) into (2.62) and balancing harmonics, one obtains recurrence relations linking A_n, B_n and $A_{n\pm2}$, $B_{n\pm2}$. Truncating these equations at successive n gives an increasingly accurate form of the stability boundary. In this way, one obtains [89] for $\Omega = 0.75$, $A_1 = 21.48$, $A_3 = 17.0826$, $A_5 = 17.1818$, $A_7 = 17.1805$.

Both the Lindsted-Poincare and the Floquet methods give very similar results for the stability boundary (Fig. 2.20). The regions of chaotic solutions obtained by the numerical solutions of Eq. (2.61) are shown in this figure [88]. The domain S_R located beyond the right-hand boundary of the large $V-$shaped domain describes the solutions $\phi = d\phi/dt = 0$ for

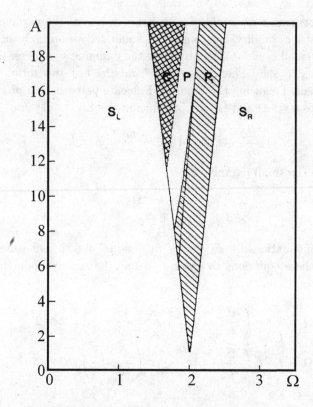

Fig. 2.20 Phase diagram for parametrically driven pendulum for $\alpha = 0.055$. Regions: P_1) trajectories with period 1, P) multiperiodic trajectories, C) chaotic trajectories. Reprinted with permission from [88]. Copyright (1989) by the American Physical Society.

all initial conditions. The domain S_L to the left of the V-shaped domain is related to different periodic and rotating solutions. The domains S_R and S_L are separated by a wedge-shaped domain subdivided into three distinct parts P_1, P and C. The first two contain oscillating and rotating solutions, respectively. Chaotic solutions are contained within domain C which is permeated by non-chaotic solutions as well.

Multiplying the sum in (2.65) by the factor $\exp(\mu t)$ gives the chaotic transient time μ^{-1} as the function of the proximity to the stability boundary [89]. The Floquet multiplier μ^{-1} has been taken [89] as a measure of the mean transient lifetime T, which is scaled as $T \approx |\,A - A_C\,|^{-1}$, where A_C defines the chaotic boundary. The numerical solution of Eq. (2.61) with $\alpha = (1.50)^{-1}$ and $A_C = 17.2$ leads to similar results [90].

2.5 Periodic and constant forces

The addition of a constant torque a to Eq. (2.1),

$$\frac{d^2\phi}{dt^2} + \alpha\frac{d\phi}{dt} + \sin\phi = a + A\sin(\Omega t) \qquad (2.66)$$

is particularly important for the analysis of Josephson junctions with applied dc and rf currents. The road to chaos was found [91] to be different for the regions with increasing values of $\langle d\phi/dt\rangle$ ("running" solutions) and for those with constant values of $\langle d\phi/dt\rangle$ ("locked" solutions). In the former case, the transition to chaos goes through the intermittency scenario, whereas in the latter case, the system exhibits period-doubling bifurcations.

The voltage-current graph ($\langle d\phi/dt\rangle$ versus a) of a Josephson junction has been analyzed [40] for different values of the damping constant α. For $A = 0$ and $a \gg 1$, the angular velocity $\langle d\phi/dt\rangle$ is very large, and one can neglect the nonlinear sinusoidal term in (2.66). For this case, the pendulum reaches the asymptotic value $\langle d\phi/dt\rangle = a/\alpha$ in a time of order α^{-1}. For $\alpha \leq 1$, the voltage-current graph exhibits hysteresis which allows a simple physical explanation [40]. Indeed, when the bias a increases from zero to unity due to the inertial term, the pendulum starts to rotate, approaching the limiting value $\langle d\phi/dt\rangle = a/\alpha$. On the other hand, upon decreasing a, the inertia causes the pendulum to continue rotating even for $a < 1$, until reaching the critical value a_{cr}, where the angular velocity vanishes, and the pendulum relaxes toward the equilibrium position. For $\alpha \to 0$, $a_{cr} \approx \sqrt{2}\alpha$. No fundamental difference occurs when $A \neq 0$, i.e., when both constant and periodic torques are present.

A comprehensive analysis has been performed [92] of Eq. (2.66), written in a slightly different form

$$\beta\frac{d^2\phi}{dt^2} + \frac{d\phi}{dt} + \sin\phi = a + A\sin(\Omega t) \qquad (2.67)$$

Equation (2.67) has been investigated numerically for $\Omega = 0.01$ over a wide range of the three other parameters, $1 < \beta < 20.000$; $0 < a < 1$; $0.1 < A < 200$. The onset of chaos was detected by studying the Poincare sections, Lyapunov exponents and power spectra. Chaos does not occur [92] when any of the following three criteria is not satisfied, namely, $\beta \ll 1$, $\Omega \gg \beta^{-1/2}$ or $A \ll 1$. The physical explanation of the first criterion is that there is no chaos for $\beta = 0$, and it appears only for β not too small. The second criterion originates from the requirement that the driving frequency

must be close to the resonant frequencies of the pendulum, and the third
criterion matches the requirements for a system to be nonlinear "enough."

2.5.1 Melnikov method

We use the Melnikov method of analysis for Eq. (2.66), which allows one
to predict the regions in the parameter plane where chaos occurs. One
rewrites Eq. (2.66) as two first-order differential equations,

$$\frac{d\phi}{dt} = y; \qquad \frac{dy}{dt} = -\sin\phi + \varepsilon\left[a - \alpha y + A\sin(\Omega t)\right] \qquad (2.68)$$

which, for $\varepsilon = 1$, reduce to Eq. (2.66), and for $\varepsilon = 0$, represent the
unperturbed system,

$$\frac{d\phi}{dt} = y; \qquad \frac{dy}{dt} = -\sin\phi, \qquad (2.69)$$

which has the heteroclinic orbit

$$\phi_h(t) = \pm 2\arctan(\sinh t); \qquad y_h(t) = \pm 2\operatorname{sech} t \qquad (2.70)$$

Inserting (2.70) into the Melnikov integral

$$M(t_0) = \int_{-\infty}^{\infty} y_h(t - t_0)\left[a + A\sin(\Omega t) - \alpha y_h(t - t_0)\right]dt, \qquad (2.71)$$

yields [66],

$$M(t_0) = \pm 2\pi a - 8\alpha \pm 2\pi A\operatorname{sech}(\pi\Omega/2)\sin(\Omega t_0) \qquad (2.72)$$

Equation (2.72) gives the Melnikov criterion for the onset of chaos,

$$A \geq \left|\pm a + \frac{4\alpha}{\pi}\left|\cosh(\pi\Omega/2)\right|\right| \qquad (2.73)$$

Numerical estimates of the Melnikov criterion [93] are shown in Fig. 2.21,
where the "Melnikov ratio" A/α is plotted as a function of Ω, and the points
give the lower boundary for the onset of chaos.

Comparing this result with the numerical solutions of Eq. (2.66) shows
[93] the following:

1. The threshold for chaos increases rapidly for $\Omega \lesssim 0.1$ and $\Omega \geq 1.2$.
These results agree with those obtained in [92] and [94], respectively.

2. The minimal value of the threshold is located near $\Omega \approx 0.6$, which
agrees with the result obtained for $a = 0$ [43].

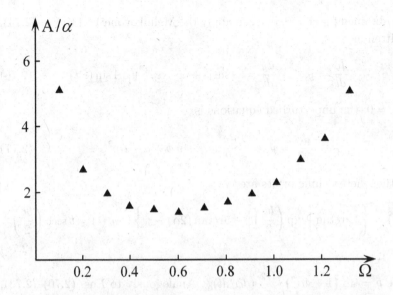

Fig. 2.21 Numerical estimates of the Melnikov ratio A/α for a pendulum subject to a constant force $a = 0.2$ and a periodic force of amplitude A (see text for explanation). Reprinted from [93], Copyright (1987), with permission from Elsevier.

3. The values of the parameters that define the onset of chaos obtained by numerical calculations are in close agreement with those obtained from the Melnikov analysis. Indeed, for $\alpha = 1/\sqrt{10}$, $\Omega = 0.01\sqrt{10}$ and $a = 0.1$, the numerical calculation gives [92] $A_{cr} = 5$, whereas the Melnikov criterion gives $A_{cr} = 5.12$. For $\alpha = 1/\sqrt{15}$, $\Omega = 0.019$, $a = 0.1$, the results of numerical calculation [94] is $A_{cr} = 1.1$, whereas the criterion (2.73) gives $A_{cr} = 1.8$.

The Melnikov method has been applied to the study of the dynamic behavior of pendulum with quadratic damping,

$$\frac{d^2\phi}{dt^2} + \alpha \left(\frac{d\phi}{dt}\right)^2 + \sin\phi = a + A\sin(\Omega t) \tag{2.74}$$

For $A = 0$, Eq. (2.74) has an exact steady-state solution,

$$\left(\frac{d\phi}{dt}\right)^2 = \frac{a\sqrt{1 + 4\alpha^2} + 2\alpha\cos\left[\phi + \arctan\left(2\alpha\right)\right]}{\alpha\sqrt{(1 + 4\alpha^2)}} \tag{2.75}$$

which gives $\phi\left(t\right)$ expressed in elliptic functions [66]. Since this solution has been obtained for $A = 0$, one can apply the Melnikov method to Eq. (2.74), rewritten as

$$\frac{d\phi}{dt} = y; \qquad \frac{dy}{dt} = -\sin\phi + a - \alpha y^2 + \varepsilon A \sin(\Omega t) \qquad (2.76)$$

For $\varepsilon = 0$, the unperturbed equations is

$$\frac{d\phi}{dt} = y; \qquad \frac{dy}{dt} = -\sin\phi + a - \alpha y^2 \qquad (2.77)$$

and their heteroclinic orbits are

$$\phi_h(t) = \left\{ 4\arctan\left[\exp\left(\frac{bt}{2}\right)\right] - \arctan(2\alpha) - \pi \right\}; \quad y_h(t) = b\,\text{sech}\left(\frac{bt}{2}\right) \qquad (2.78)$$

where $b = 4\alpha\left(1 + 4\alpha^2\right)^{-1/2}(d\phi/dt)_0$. Analogously to Eqs. (2.70)–(2.73), one obtains the Melnikov function and the Melnikov criterion for the onset of chaos,

$$A > \frac{[a - 2\alpha(1 + 4\alpha^2)^{-1/2}]\sinh(2\pi\alpha)}{\alpha b\sqrt{F_1^2 + F_2^2}} \qquad (2.79)$$

where F_1 and F_2 are integrals [95].

Note that the criterion (2.73), obtained from Eq. (2.66), was based on the unperturbed Eqs. (2.69), whereas for the quadratic damping Eq. (2.74), we used the unperturbed Eq. (2.77). In the latter case, one could also use Eq. (2.69) as the appropriate unperturbed equation by considering the $a - \alpha y^2$ terms as being of order ε [95]. The only difference will be the forms of the homoclinic orbits for the unperturbed Eqs. (2.69) and (2.77). Both Melnikov criteria (2.72) and (2.79) will describe the onset of chaos for the quadratically damped pendulum. Thus, one can compare the results of the Melnikov analysis for different perturbation methods. It is evident that for small damping α and a small constant term a, there is no significant difference between these two criteria. However, for intermediate damping and a large constant force, the results are quite different. Due to the different division of the equation into the unperturbed part and the perturbation, for comparatively large α, the criterion (2.79) is close to the results of the numerical calculations, whereas the criterion (2.73) is useless [95].

2.6 Parametric and constant forces

In Section 2.4, we considered a pendulum subject to a parametric periodic force, which means that the potential function, $V = (1 + A\sin(\Omega t))\cos\phi$, is symmetric with respect to replacing ϕ by $-\phi$. Such an idealization is permissible in theoretical calculations. However, in experimental arrangements, the assumption of symmetry is only approximate, and in many cases, the bias term, such as the dc current in a Josephson junction, or a constant wind load in cargo loading [96], has to be added to the dynamic equation. This leads to the following equation of motion

$$\frac{d^2\phi}{dt^2} + \alpha\frac{d\phi}{dt} + (1 + A\sin(\Omega t))\sin\phi = a \qquad (2.80)$$

The effect of the bias term a on the behavior of an unexcited pendulum ($A = \alpha = 0$) has been considered in Section 1.4. A similar influence occurs in the more general case (2.80) which manifests itself in the decrease of the onset of rotating motions and in the change of the number and the types of attractors. Moreover, the stability of the parametrically excited pendulum disappears if the bias term exceeds a critical value.

In the series of articles, S.-Y. Kim and collaborators [97] analyzed the pendulum equation of the form

$$\frac{d^2x}{dt^2} + 2\pi\alpha\frac{dx}{dt} + 2\pi\left[\omega_0^2 + A\cos\left(2\pi t\right)\right]\sin\left(2\pi x\right) = 0 \qquad (2.81)$$

Upon changing the variables (x by $\phi/2\pi$ and t by $t/2\pi$), Eq. (2.81) becomes

$$\frac{d^2\phi}{dt^2} + \alpha\frac{d\phi}{dt} + \left[\omega_0^2 + A\cos\left(t\right)\right]\sin\phi = 0 \qquad (2.82)$$

Equation (2.82) has three control parameters: the amplitude A of an external field, the damping coefficient α, and the frequency ω_0 of the linearized pendulum. The solutions of Eq. (2.82) were analyzed [97] for different A and ω_0 at $\alpha = 0.1$. For the force-free pendulum ($A = 0$), the normal fixed point $\phi = 0$, $d\phi/dt = 0$ is stable, whereas the inverted fixed point $\phi = \pi$, $d\phi/dt = 0$ is unstable. An increase of A is accompanied by the restabilization of the fixed points and the transition to chaos, depending on the value of the parameter ω_0. We present here the results of the numerical solution of Eq. (2.82), which agree with experiment for a parametrically forced magnetic pendulum [97]. As A increases, the normal fixed point becomes unstable. Depending on the value of ω_0, the loss of stability of the normal fixed point occurs either by a period-doubling bifurcation or by a symmetry-breaking pitchfork bifurcation. A new stable trajectory with period two appears at a critical value A_1 ($A_1 = 0.1002$ for $\omega_0 = 0.5$) via period-doubling bifurcation. As A increases, the symmetric period-2 trajectory becomes unstable and transfers to the asymmetric period-2 trajectory, accompanied by an infinite series of period-doubling bifurcations, ending at the accumulation point $A_{acc} = 0.3578$. After the period-doubling transition to chaos, a further increase of A restores the regular trajectories. In contrast to the normal fixed point, the inverted fixed point first increases its stability up to $A < A_2$, but then destabilizes for $A > A_3$, being stable in the region $A_2 < A < A_3$. Such order-chaos-order sequences alternate with a further increase of A.

We illustrate the behavior of the solution of Eq. (2.80) by two approximate analytical methods, the harmonic balance method [98] and Melnikov analysis [99], as well as by the numerical solution [98].

2.6.1   *Harmonic balance method*

For $a = 0$, the symmetric system has a zone around $\Omega = 2$ for which initial conditions close to $\phi_{t=0} = (d\phi/dt)_{t=0} = 0$ lead to oscillating solutions. Outside this "safe zone," one obtains rotating solutions. It is interesting to find the boundaries of the safe zone for the symmetry-breaking Eq. (2.80). In the presence of the nonsymmetric a-term, Eq. (2.80) does not have $\phi_{t=0} = (d\phi/dt)_{t=0} = 0$ as an equilibrium solution, which leads to a decrease of the safe zone. This is clear for the nonsymmetric "washboard" form of the potential function $V = (1 + A\sin(\Omega t))\cos x - ax$, which decreases the barrier of the transition from oscillating to rotating motion. One can obtain an approximate solution from the harmonic balance method [98; 101]. Assume that the approximate solution of Eq (2.80) has the following form,

$$\phi(t) = \phi_0 + c\cos\left[2\left(\Omega t + \delta\right)\right] + \eta \qquad (2.83)$$

where δ is the phase angle and η is a small amplitude such that only terms of order η are taken into account in the calculations. Inserting (2.83) into (2.80) and equating the coefficients of the cosine, the sine and the constant, one obtains three equations. The solutions of these equations are non-divergent if [98]

$$\phi_0 = (-3a)^{1/3}; \qquad c = \left[\frac{8 - 4\left(9a^2\right)^{1/3}}{2 + A\cos\delta}\right]^{1/2} \qquad (2.84)$$

and the following approximate expression for the amplitude A_{cr}, which defines the boundary of safe zone,

$$A_{cr}^2 = \frac{\left[1 - \dfrac{\Omega^2}{4} - \dfrac{\phi_0^2}{2} - \dfrac{c_0^2}{8}\right] + \left(\dfrac{c\Omega}{2}\right)^2}{\left(\dfrac{\phi_0^2}{4} - \dfrac{\phi_0}{2c_0} + \dfrac{c_0^2}{12} - \dfrac{1}{2}\right)^2} \qquad (2.85)$$

where $c_0 \equiv c(\delta = 0)$. For $a = 0$, this result reduces to that obtained for the symmetric parametrically driven pendulum [102]. It is evident that along the boundary of the safe zone, the presence of the constant force in Eq. (2.80) leads to a shift of the onset of period-doubling trajectories and of deterministic chaos.

2.6.2 *Heteroclinic and homoclinic trajectories*

As shown in Section 1.4, the addition of an excitation a-term to Eq. (2.1) transforms the heteroclinic orbit, linking two saddle points, into a homoclinic orbit going from a saddle point x_0 to a turning point x_1. Moreover, the presence of $A\cos(\Omega t)$ poses difficulties for computing the Melnikov function. To overcome this problem, one changes the time variable t into a function of the state variable x on the homoclinic orbit [103]. This results in the following Melnikov function

$$M(t_0) = -\alpha\mu_1 + A\mu_2 \sin(\Omega t_0) \qquad (2.86)$$

where

$$\mu_1 = \mp 2 \int_{x_1}^{x_0} dx \sqrt{2(H + \cos x - ax)}; \qquad (2.87)$$

$$H = \frac{1}{2}\left(\frac{d\phi}{dt}\right)^2 - \cos\phi - a\phi$$

and

$$\mu_2 = \pm 2 \int_{x_1}^{x_0} dx \sin x \sin\left[\Omega \int_{x_1}^{x_0} \frac{dy}{\sqrt{2(H + \cos y - ay)}}\right] \qquad (2.88)$$

Calculating the integrals in (2.87) and (2.88), and inserting μ_1 and μ_2 into the Melnikov function (2.86), one obtains [99] the homoclinic bifur-

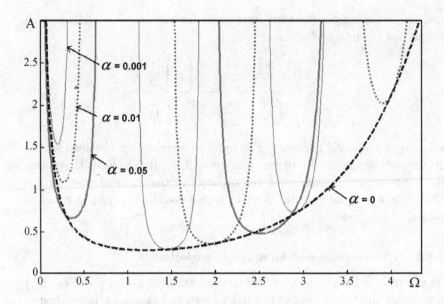

Fig. 2.22   Homoclinic bifurcation curves for a pendulum subject to constant and parametric periodic forces for different values of the constant force. Reprinted from [99],
Copyright (2008), with permission from Elsevier.

cation curve shown in Fig. 2.22 for different values of a in the parametric
plane (A, Ω). One sees from this figure that as a increases, the region which
leads to chaos is enlarged, i.e., chaotic motion covers a larger portion of the
parametric plane as the bias a-term increases.

2.6.3   *Numerical calculations*

The numerical solution of Eq. (2.80) has been obtained [98] for $\Omega = 2.1$ and
$\alpha = 0.1$. The results are shown in Fig. 2.23, where we display the amplitude
of an external force A needed for the transition from oscillating to rotating
trajectories, as a function of the bias force a. The shaded area represents the
region of oscillating trajectories ("safe zone"). The points labeled I, II and
III give, respectively, the first period-doubling, the period-doubling that
initiates the sequence of the period-doubling cascade, and the supercritical
period-doubling just before transition to rotating trajectories. The points
S_A and S_B label the coexistence of the stable-unstable pair of subharmonic
orbits for $a = 0.0162$ and $a = 0.02$.

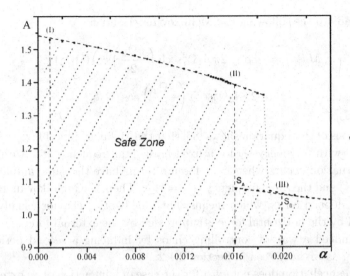

Fig. 2.23 Critical value of the amplitude of a parametric periodic force which defines the transition from oscillating to rotating trajectories, as a function of a constant bias force for a pendulum with $\Omega = 2.1$ and $\alpha = 0.1$ (see text for explanation). Reprinted from [98], Copyright (2006), with permission from Elsevier.

2.7 External and parametric periodic forces

As shown in Sections 2.1 and 2.3, chaotic trajectories can be obtained by adding a periodic force to the pendulum equation of motion, either additive (external force) or multiplicative (vertical oscillations of the suspension point). Such trajectories occur for specific regions of the control parameters. For some practical applications, however, one needs to control or even depress chaos by an additional external factor. The constant force may play the role of such a factor, both for an additive and for a multiplicative periodic force. These two cases have been considered in Sections 2.1 and 2.3. The case of two external periodic forces was the subject of Section 2.1.2. Still another possibility of an external influence on chaos is given by both additive and multiplicative periodic forces. Therefore, we shall analyze a pendulum subject to both an external force and to horizontal oscillations of the suspension point. The equation of motion is

$$\frac{d^2\phi}{dt^2} + \alpha\frac{d\phi}{dt} + A_1 \cos\left(\Omega_1 t\right) \cos\phi + \sin\phi = A_2 \sin(\Omega_2 t) \qquad (2.89)$$

which leads to the following the Melnikov function [86]

$$M(t_0) = -8\alpha - 2\pi A_1 \Omega_1^2 \text{sech}\left(\frac{\pi\Omega_1}{2}\right)\cos(\Omega_1 t_0)$$

$$+2\pi A_2 \Omega_2^2 \text{sech}\left(\frac{\pi\Omega_2}{2}\right)\cos(\Omega_2 t_0) \qquad (2.90)$$

Analysis of this equation shows [86] that the Melnikov function decreases smoothly with frequency Ω_2 for a fixed horizontal frequency Ω_1. An interesting situation occurs when both forcing terms have the same amplitude ($A_1 = A_2$) and the same frequency ($\Omega_1 = \Omega_2$). Then, the Melnikov function does not depend on t_0, which means that the chaotic behavior usually introduced by the horizontal force is suppressed by the external forcing term. This is indeed a way of controlling chaos by inducing a second periodic force [49], as we have seen in Section 2.1.2.

Such cancellation does not exist for the case of joint action of an external force and the vertical oscillations of the suspension point. In this case, the equation of motion is

$$\frac{d^2\phi}{dt^2} + \alpha\frac{d\phi}{dt} + [1 + A_1\cos(\Omega_1 t)]\sin\phi = A_2\sin(\Omega_2 t) \qquad (2.91)$$

which leads to the following Melnikov function

$$M(t_0) = -8\alpha - 2\pi A_1 \Omega_1^2 \text{csch}\left(\frac{\pi\Omega_1}{2}\right)\sin(\Omega_1 t_0)$$

$$+2\pi A_2 \text{sech}\left(\frac{\pi\Omega_2}{2}\right)\cos(\Omega_2 t_0), \qquad (2.92)$$

with non-vanishing time-dependence terms.

Slightly more complicated is the equation of motion containing the spring constant β,

$$\frac{d^2\phi}{dt^2} + \alpha\frac{d\phi}{dt} + \beta\phi + [1 + A_1\cos(\Omega_1 t)]\sin\phi = A_2\sin(\Omega_2 t) \qquad (2.93)$$

which has been analyzed [100]. The chaotic behavior induced by the parametric excitation for amplitude A and frequency Ω can be reduced to periodic motion by adjusting the control parameters α, β, and Ω. Extensive numerical studies of Eq. (2.93) have shown that varying β in the range $0.01 < \beta < 1$ for fixed values of parameters $\alpha = 0.2$, $A_1 = A_2 = 4$, $\Omega_1 = 4$

and $\Omega_2 = 2$, leads to the onset of chaos at $\beta \approx 0.538$ and $\beta \approx 0.507$ as β decreases. The reverse transition from chaotic behavior to periodic behavior occurs as β increases.

The more general case, with three harmonic forces are acting on the pendulum, is described by the following equation

$$\frac{d^2\phi}{dt^2} + \alpha\frac{d\phi}{dt} + \omega_0^2\phi + A_1\cos(\Omega_1 t)\cos\phi + A_2\cos(\Omega_2 t)\sin\phi = A_3\sin(\Omega_3 t) \tag{2.94}$$

For simplicity, we restrict ourselves to the case where $A_1 = A_2 = A_3 \equiv A$ and $\Omega_1 = \Omega_2 = \Omega_3 \equiv \Omega$ [86]. The Melnikov function $M(t_0)$ is

$$M(t_0) = -8\alpha\omega_0 + 2\pi A\left[\left(1 - \frac{\Omega^2}{\omega_0^2}\right)\operatorname{sech}\left(\frac{\pi\Omega}{2\omega_0}\right)\cos(\Omega t_0)\right.$$

$$\left. + \frac{\Omega^2}{\omega_0^2}\operatorname{csch}\left(\frac{\pi\Omega}{2\omega_0}\right)\sin(\Omega t_0)\right] \tag{2.95}$$

The time $t_{0,\max}$ which maximizes the Melnikov function (2.95) is

$$\tan t_{0,\max} = \frac{\Omega^2}{\omega_0^2 - \Omega^2}\coth\left(\frac{\pi\Omega}{2\omega_0}\right) \tag{2.96}$$

and the Melnikov function at this time is

$$M(t_{0,\max}) = -8\omega_0\alpha + \frac{2\pi A}{\omega_0^2}\left[\Omega^4\operatorname{csch}^2(\pi\Omega/2\omega_0)\right.$$

$$\left. + \left(\omega_0^2 - \Omega^2\right)^2\operatorname{sech}^2(\pi\Omega/2\omega_0)\right]^{1/2} \tag{2.97}$$

The zero of $M(t_{0,\max})$ defines the parameters close to the onset of chaos. The Melnikov criterion for the onset of chaos is,

$$A > \frac{4\alpha\omega_0^3}{\pi\left[\Omega^4\operatorname{csch}^2(\pi\Omega/2\omega_0) + (\omega_0^2 - \Omega^2)^2\operatorname{sech}^2(\pi\Omega/2\omega_0)\right]^{1/2}} \tag{2.98}$$

Equation (2.98) shows that for $\omega_0 = 1.5$, the Melnikov ratio A/α increases with external frequency Ω, and exhibits a local maximum for Ω close to ω_0. This implies that in the resonance region, one has to increase the strength of the perturbation in order to produce chaotic behavior.

Chapter 3

Pendulum subject to a Random Force

3.1 Noise

3.1.1 *White noise and colored noise*

We will consider noise $\xi(t)$ having $\langle \xi(t) \rangle = 0$ and the correlator

$$\langle \xi(t_1) \xi(t_2) \rangle = r(|t_1 - t_2|) \equiv r(z). \tag{3.1}$$

Two integrals of (3.1) characterize fluctuations: the strength of the noise D,

$$D = \frac{1}{2} \int_0^\infty \langle \xi(t) \xi(t+z) \rangle \, dz, \tag{3.2}$$

and the correlation time τ,

$$\tau = \frac{1}{D} \int_0^\infty z \langle \xi(t) \xi(t+z) \rangle \, dz. \tag{3.3}$$

Traditionally, one considers two different forms of noise, white noise and colored noise. For white noise, the function $r(|t_1 - t_2|)$ is a delta-function,

$$\langle \xi(t_1) \xi(t_2) \rangle = 2D\delta(t - t_1). \tag{3.4}$$

The name "white" noise derives from the fact that the Fourier transform of (3.4) is "white", that is, constant without any characteristic frequency. Equation (3.4) implies that noise $\xi(t_1)$ and noise $\xi(t_2)$ are statistically independent, no matter how near t_1 is to t_2. This extreme assumption, which leads to the non-physical infinite value of $\langle \xi^2(t) \rangle$ given by (3.4), does not imply that the correlation time τ is zero, as was assumed in (3.4), but rather that it is smaller than all other characteristic times in the problem.

We will return to this subject in Section 3.1.3, when examining the Fokker-Planck equation for multiplicative noise.

All sources of non-white noise are called colored noise. A widely used form of noise is the Ornstein-Uhlenbeck exponentially correlated noise,

$$\langle \xi(t)\,\xi(t_1) \rangle = \sigma^2 \exp\left[-\lambda\,|t - t_1|\right], \tag{3.5}$$

or, equivalently,

$$\langle \xi(t)\,\xi(t_1) \rangle = \frac{D}{\tau} \exp\left(-\frac{|t - t_1|}{\tau}\right). \tag{3.6}$$

White noise (3.4) is characterized by its strength D, whereas Ornstein-Uhlenbeck noise is characterized by two parameters λ and σ^2, or τ and D. The transition from Ornstein-Uhlenbeck noise to white noise (3.4) occurs in the limit $\tau \to 0$ in (3.6), or when $\sigma^2 \to \infty$ and $\lambda \to \infty$ in (3.5) such that $\sigma^2/\lambda = 2D$.

A slightly generalized form of Ornstein-Uhlenbeck noise is narrow-band colored noise with a correlator of the form,

$$\langle \xi(t)\,\xi(t_1) \rangle = \sigma^2 \exp(-\lambda\,|t - t_1|) \cos\left(\Omega\,|t - t_1|\right). \tag{3.7}$$

There are different forms of colored noise, one of which will be described in the next section.

3.1.2 *Dichotomous noise*

A special type of colored noise is symmetric dichotomous noise (random telegraph signal) for which the random variable $\xi(t)$ may equal $\pm\sigma$ with mean waiting time $(\lambda/2)^{-1}$ in each of these two states. Like Ornstein-Uhlenbeck noise, dichotomous noise is characterized by the correlators (3.5) and (3.6).

We will use the Shapiro-Loginov procedure [104] for dealing with higher-order correlations. For exponentially correlated noise, this yields

$$\frac{d}{dt}\langle \xi \cdot g \rangle = \left\langle \xi\frac{dg}{dt} \right\rangle - \lambda \langle \xi \cdot g \rangle, \tag{3.8}$$

where g is some function of the noise, $g = g\{\xi\}$. If $dg/dt = B\xi$, then Eq. (3.8) becomes

$$\frac{d}{dt}\langle \xi \cdot g \rangle = B\left\langle \xi^2 \right\rangle - \lambda \langle \xi \cdot g \rangle, \tag{3.9}$$

and for white noise ($\xi^2 \to \infty$ and $\lambda \to \infty$, with $\xi^2/\lambda = 2D$),

$$\langle \xi \cdot g \rangle = 2BD. \tag{3.10}$$

3.1.3 *Langevin and Fokker-Planck equations*

Noise was introduced into differential equations by Einstein, Smoluchowski and Langevin when they considered the molecular-kinetic theory of Brownian motion. They assumed that the total force acting on the Brownian particle can be decomposed into a systematic force (viscous friction proportional to velocity, $f = -\gamma v$) and a fluctuating force $\xi(t)$ exerted on the Brownian particle by the molecules of the surrounding medium. The fluctuating force derives from the molecular collisions on a Brownian particle from all sides resulting in random motion. The motion of a Brownian particle of mass m is described by the so-called Langevin equation

$$m\frac{dv}{dt} = -\gamma v + \xi(t) \tag{3.11}$$

The stochastic equation (3.11) describes the motion of an individual Brownian particle. The random force $\xi(t)$ in this equation causes the solution $v(t)$ to be random as well. One can consider an ensemble of Brownian particles and ask how many particles in this ensemble have velocities in the interval $(v, v + dv)$ at time t, which defines the probability function $P(v, t)\,dv$. The deterministic equation for $P(v, t)$ is called the Fokker-Planck equation, which has the following form for white noise [105],

$$\frac{\partial P(v, t)}{\partial t} = \frac{\partial}{\partial v}(\gamma v P) + D\frac{\partial^2 P}{\partial v^2}. \tag{3.12}$$

In the general case, in which the equation of motion $dx/dt = f(x)$ has a nonlinear function $f(x)$, the Langevin equation is

$$\frac{dx}{dt} = f(x) + \xi(t) \tag{3.13}$$

with the appropriate Fokker-Planck equation being

$$\frac{\partial P(x, t)}{\partial t} = -\frac{\partial}{\partial v}[f(x)P] + D\frac{\partial^2 P}{\partial v^2} \tag{3.14}$$

Thus far, we have considered additive noise which describes an internal noise, say, thermal noise. However, there are also fluctuations of the surrounding medium (external fluctuations) which enter the equations as multiplicative noise,

$$\frac{dx}{dt} = f(x) + g(x)\xi(t). \tag{3.15}$$

The appropriate Fokker-Planck equation then has the form [105]

$$\frac{\partial P(x,t)}{\partial t} = -\frac{\partial}{\partial x}[f(x)P] + D\frac{\partial}{\partial x}g(x)\frac{\partial}{\partial x}g(x)P. \tag{3.16}$$

We shall not discuss the Ito-Stratonovich dilemma [105] connected with Eq. (3.16).

The preceding discussion was restricted to first-order stochastic differential equations. Higher-order differential equations can always be written as a system of first-order equations, and the appropriate Fokker-Planck equation for such a system has the following form

$$\frac{\partial P(x,t)}{\partial t} = -\sum_i \frac{\partial}{\partial x_i}[f_i(x)P] + \frac{1}{2}\sum_{i,j}\frac{\partial^2}{\partial x_i \partial x_j}[g_{ij}(x)P] \tag{3.17}$$

for any functions $f_i(x)$ and $g_{ij}(x)$. The linearized form of (3.17) is

$$\frac{\partial P(x,t)}{\partial t} = -\sum_{i,j} f_{ij}\frac{\partial}{\partial x_i}(x_jP) + \frac{1}{2}\sum_{i,j}g_{ij}\frac{\partial^2 P}{\partial x_i \partial x_j} \tag{3.18}$$

where f_{ij} and g_{ij} are constant matrices.

For colored noise, there is no rigorous way to find the Fokker-Planck equation that corresponds to the Langevin equations (3.13) and (3.15). Therefore, one has to use various approximations [19].

One can illustrate [106] the importance of noise in deterministic differential equations by the simple example of the Mathieu equation supplemented by white noise $\xi(t)$

$$\frac{d^2\phi}{dt^2} + (\alpha - 2\beta\cos 2t)\phi = \xi(t) \tag{3.19}$$

The solutions of Eq. (3.19) in the absence of noise are very sensitive to the parameters α and β, which determine regimes in which the solutions can be periodic, damped or divergent. To obtain the Fokker-Planck equation corresponding to the Langevin equation (3.19), we decompose this second-order differential equation into the two first-order equations

$$\frac{d\phi}{dt} = \Omega; \qquad \frac{d\Omega}{dt} = -(\alpha - 2\beta\cos 2t)\phi + \xi(t) \tag{3.20}$$

The Fokker-Planck equation (3.17) for the distribution function $P(\phi, \Omega, t)$ then takes the form

$$\frac{\partial P}{\partial t} = D\frac{\partial^2 P}{\partial \Omega^2} - \Omega\frac{\partial P}{\partial \phi} + (\alpha - 2\beta\cos 2t)\phi\frac{\partial P}{\partial \Omega} \qquad (3.21)$$

with initial conditions $P(\phi, \Omega, 0) = \delta(\phi - \phi_0)\delta(\Omega - \Omega_0)$. Equation (3.21) can easily be solved by Fourier analysis to obtain the following equation for the variance $\sigma^2 \equiv \langle\phi^2\rangle - \langle\phi\rangle^2$,

$$\frac{d^3(\sigma^2)}{dt^3} + 4(\alpha - 2\beta\cos 2t)\frac{d(\sigma^2)}{dt} + (8\beta\sin 2t)\sigma^2 = 8D \qquad (3.22)$$

The solution of Eq. (3.22) has the same qualitative features as the solution of Eq. (3.19). However, for sufficiently large values of β, the variance does not exhibit the expected linear diffusion dependence but increases exponentially with time.

For this simple case of a linear differential equation, one can obtain an exact solution. However, the equation of motion of the pendulum is nonlinear. We will see from the approximate calculations and numerical solutions that the existence of noise modifies the equilibrium and dynamic properties of the pendulum in fundamental way.

3.2 External random force

There are various simplified forms of the pendulum equation with a random force. We begin with the following equation,

$$\frac{d^2\phi}{dt^2} + \alpha\frac{d\phi}{dt} + \sin\phi = \xi(t) \qquad (3.23)$$

which describes a damped Brownian particle moving in a periodic potential. According to the fluctuation-dissipation theorem for a stationary state, a gain of energy entering the system is exactly compensated by the energy loss to the reservoir, which gives

$$\langle\xi(t)\xi(0)\rangle = 2\alpha\kappa_B T\delta(t) \qquad (3.24)$$

where κ_B is the Boltzmann constant. The velocity-velocity autocorrelation function gives the frequency-dependent mobility $\mu(\omega, T)$,

$$\mu(\omega, T) = \frac{1}{\kappa_B T}\int_0^\infty dt\left\langle\frac{d\phi}{dt}(t)\frac{d\phi}{dt}(0)\right\rangle\exp(i\omega t) \qquad (3.25)$$

Like other pendulum equations, Eq. (3.23) usually allows only a numerical solution [19]. However, for special cases, can one obtain analytical results for $\mu(0,T)$. For the non-damped case $(\alpha = 0)$ [107],

$$\mu(0,T) = \pi(2\pi\kappa T)^{-1/2}\frac{I_0(v) + I_1(v)}{I_0^2(v)}\exp(-v) \qquad (3.26)$$

Here $v = \Delta E/\kappa_B T$, where ΔE is the barrier height, and $I_n(v)$ is the Bessel function of nth order. In the limit of large damping [108], neglecting the second derivative in (3.23) yields

$$\mu(0,T) = \frac{1}{\alpha[I_0^2(v)]} \qquad (3.27)$$

The method of continued fractions [19] can be used to obtain an approximate expression for the mobility

$$\mu(\omega) = \frac{1}{-i\omega + \alpha - I_1(v)/i\omega I_0(v)} \qquad (3.28)$$

which is in good agreement with the numerical calculations [109].

3.3   Constant and random forces

For this case, the equation of motion has the following form,

$$\frac{d^2\phi}{dt^2} + \alpha\frac{d\phi}{dt} = a - \sin\phi + \xi(t) \qquad (3.29)$$

where $\xi(t)$ is the white noise. This equation cannot be solved analytically, but one may present a qualitative analysis. Equation (3.29) corresponds to motion in the washboard potential, shown in Fig. 3.1,

$$U(\phi) = -a\phi - (1 - \cos\phi). \qquad (3.30)$$

Neglecting the second derivative and the random force $\xi(t)$ in Eq. (3.29), one obtains two types of solutions: locked-in for $a < 1$, and running for $a > 1$. A similar situation occurs [110] for the full equation (3.29). The system switches randomly between a locked state with zero average velocity, $\langle d\phi/dt \rangle = 0$, and a running state with $\langle d\phi/dt \rangle \neq 0$. For $a < 1$, the transition from the locked state to the running state occurs when the value of a approaches unity, whereas for $a > 1$ the opposite transition takes place when the value of a decreases to $4\alpha/\pi$. For large values of α, say, $\alpha > \pi/4$,

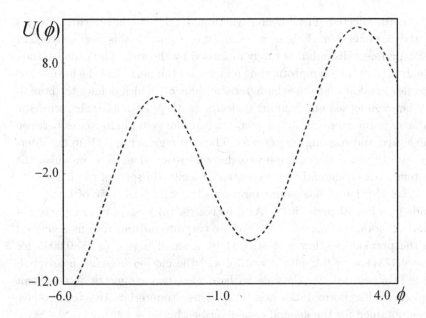

Fig. 3.1   Washboard potential $U(\phi) = a_0\phi - b_0 \cos\phi$.

one may neglect the second derivative in Eq. (3.29). In this overdamped case the two boundary values coincide, leading to the criterion $a \gtrless 1$.

For weak noise, the stationary dynamics of $\dot{\phi}(t)$ is controlled by the critical value of $a_{cr} = (2 + \sqrt{2})\alpha$. For $a < a_{cr}$, the system is trapped in small oscillations near the downward position, whereas for $a > a_{cr}$, the pendulum performs rotations.

The results are different for large and small damping constant α [19]. For large α, one can neglect the inertial term $d^2\phi/dt^2$. (This overdamped pendulum will be considered in Section 3.8). Because of the noise, pendulum will not stay in one locked-in position, but will sometimes be transformed to the other locked-in state. For a small damping constant, inertial effects are important and the pendulum will not be trapped in a locked-in position, but may be transferred to and remain for some time in the running position. The existence of noise will lead to transitions between locked-in and running states.

For the running states, the average angular velocity $\langle d\phi/dt \rangle$ and the time evolution of the mean square displacement of the rotation angle $\langle \phi^2 \rangle$ are the characteristics of pendulum motion.

An unexpected phenomenon was found [111] for intermediate values of the parameter a. This is precisely the region that is very important for experimentalists, but is rarely discussed by theorists who usually study (analytically) the asymptotic time regime and (numerically) the initial time regime. In addition to the characteristic value a_{cr} which defines the boundary between locked and running trajectories, there is another characteristic value of parameter a, well below a_{cr}, where the system alternates between the locked and running trajectories. The new regime lies [111] in the intermediate region between these two characteristic values of a, in which the system moves coherently with constant velocity (in spite of the noise!).

The addition of a constant term a to the right hand side of Eq. (3.34) leads to a loss of periodicity. As a increases, the system alternatively becomes chaotic and periodic. One of the two intermittent regions described in the previous section is destroyed by a small bias a ($a = 0.0005$ for $\Omega = 0.47311$, $\alpha = 0.2$, and $A = 1.2$), and the motion becomes nearly ballistic. By contrast, for the case without bias, two symmetric intermittent regions lead to normal diffusion [112]. These counterintuitive results have been obtained for the general case described by Eq. (3.42).

The Fokker-Planck equation for the density distribution function $P(\phi, t)$, corresponding to the Langevin equation (3.23), has the following form,

$$\frac{\partial P}{\partial t} = \frac{\partial}{\partial \phi}\left(\frac{\sin\phi}{\alpha}p + D\frac{\partial P}{\partial \phi}\right) \equiv -\frac{\partial J}{\partial \phi} \qquad (3.31)$$

In the stationary state, $\partial J/\partial\phi = 0$ and the flux J is constant. Because of periodicity, the boundary condition is $P(0) = P(L)$. Using this condition and normalization, one can integrate Eq. (3.31) to obtain [113]

$$\left\langle\frac{d\phi}{dt}\right\rangle = \frac{D\left[1 - \exp\left(-aL/\alpha D\right)\right]}{\displaystyle\int_0^L \exp\left(-a\phi/\alpha D\right) I_0\left[D^{-1}\sin\left(\pi\phi/L\right)\right]d\phi} \qquad (3.32)$$

where $I_0(\phi)$ is the modified Bessel function. For weak noise $D << 1$, the inverse Bessel function takes the Arrhenius form

$$I_0^{-1}\left(D^{-1}\right) \approx \sqrt{\frac{2\pi}{D}}\exp\left(-D^{-1}\right), \qquad (3.33)$$

as required.

3.4 External periodic and random forces

3.4.1 *Two sources of noise*

The dynamic equation for a driven pendulum with noise is

$$\frac{d^2\phi}{dt^2} + \alpha\frac{d\phi}{dt} + \sin\phi = A\cos(\Omega t) + \xi(x) \qquad (3.34)$$

This equation shows that there are two sources of noise: one source follows from deterministic chaos which exists for certain sets of parameters of a nonlinear problem, and a second source that follows from the ever present fluctuations, such as thermal fluctuations. The latter is generally ignored in numerical calculations (if no additional random force is added to the dynamic equations), but is always present in experiments. The interplay between these two sources of experimentally observable noise is very complicated, and it is hard to distinguish between them [114]. Nevertheless, in some cases (three-photon parametric amplifier), it was shown [115] that the noise seen in experiments cannot arise from deterministic chaos. The existence of two types of noise presents a problem for comparing the numerical calculations with experiment. Small amounts of thermal noise can drastically change the solution of the dynamic equations, and therefore, must be taken into account in the explanation of experiments. In the absence of thermal noise, numerical calculations shows many complicated structures, such as bifurcations, periodic and random solutions, etc , which are typically washed out by thermal noise.

Another form of Eq. (2.1) consists of including a quenched disorder substrate by adding the random force $\beta\xi(x)$,

$$\frac{d^2\phi}{dt^2} + \alpha\frac{d\phi}{dt} + \sin\phi = A\cos(\Omega t) + \beta\xi(\phi) \qquad (3.35)$$

where $\xi(\phi) \in [-1,1]$ was chosen [116] as an independent, uniformly distributed random variable with no spatial correlations, corresponding to a piecewise constant force on the interval $[2n\pi, (2n+1)\pi]$, and β is the magnitude of the quenched disorder. The numerical solution of Eq. (3.35) for $\alpha = 0.2$, $A = 1.2$ and $\Omega = 0.3$ shows [116] that the diffusion remains normal, and the diffusion coefficient $\langle\phi^2(t)\rangle/2t$ increases with β for $\beta < 0.1$. At higher levels of disorder, $\beta > 1$, the diffusion coefficient approaches zero. The latter phenomenon stems from the suppression of chaotic diffusion for a large amount of disorder, which causes almost all trajectories to be localized.

3.4.2   *Fokker-Planck equation*

A comprehensive analysis has been performed [117] of the joint action of both factors capable of producing chaotic behavior (deterministic chaos and a random force). In order to transform the Langevin equation (3.34) into the appropriate Fokker-Planck equation, one introduces the following change of variables

$$x_1 = \phi; \quad x_2 = \frac{d\phi}{dt}; \quad x_3 = A\cos(\Omega t); \quad x_4 = \frac{dx_3}{dt} \qquad (3.36)$$

Equation (3.34) then takes the form

$$\frac{dx_1}{dt} = x_2; \quad \frac{dx_2}{dt} = -\alpha x_2 - \sin(x_1) + x_3 + \xi(t); \qquad (3.37)$$

$$\frac{dx_3}{dt} = x_4; \quad \frac{dx_4}{dt} = -\Omega^2 x_3$$

The Fokker-Planck equation for the distribution function $P(x_1, x_2, x_3, x_4)$ (with the initial conditions $x_{1,0}$ and $x_{2,0}$) is

$$\frac{\partial P}{\partial t} = -\frac{\partial}{\partial x_1}(x_2 P) - \frac{\partial}{\partial x_2}[(-\alpha x_2 - \sin x_1 + x_3)P] \qquad (3.38)$$

$$-\frac{\partial}{x_3}(x_4 P) - \frac{\partial}{\partial x_4}(-\Omega^2 x_3 P) + \frac{1}{2}D\frac{\partial^2 P}{dx^2}$$

Equation (3.38) has been solved numerically [117] for $\alpha = 1.0$, $\Omega = 0.25$ and $D = 0.5$ for two sets of initial conditions: $x_{1,0} = x_{2,0} = 0$ and $x_{1,0} = 0$, $x_{2,0} = 0.1$. For these values, the steady-state distribution function P shows chaotic behavior. Note that for these parameters, Eq. (3.34) without noise also shows chaotic behavior [118]. As we will see in Section 3.5, in contrast to additive noise, multiplicative noise is able to convert chaotic trajectories into non-chaotic trajectories.

3.4.3   *Ratchets*

The term "ratchet" denotes a periodic potential which is anisotropic. Ratchets produce a net flux without a driving force (bias force or external gradient). Detailed information above ratchets can be found in a recent review article [119].

The equation of motion of a particle moving in a ratchet potential is given by

$$m\frac{d^2x}{dt^2} + \gamma\frac{dx}{dt} = -\frac{dV(x)}{dx} + f\sin(\omega t) + \xi(t) \qquad (3.39)$$

where the anisotropic potential is shown in Fig. 3.2 [120],

$$V(x) = -\sin x - \frac{1}{4}\sin(2x) \qquad (3.40)$$

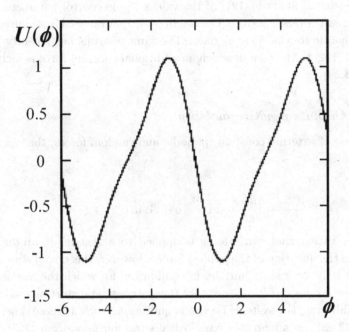

Fig. 3.2 Ratchet potential $U(\phi) = -\sin\phi - \frac{1}{4}\sin(2\phi)$. Reprinted with permission from [120]. Copyright (1996) by the American Physical Society.

In the overdamped case ($m = 0$) and in the absence of noise ($\xi = 0$), the solutions of Eq. (3.39) are either locked or running. In the latter case, the asymptotic value of the average velocity is

$$v_{qr} = \frac{x(t + rT) - x(t)}{rT} = \frac{q}{r}\omega \qquad (3.41)$$

for integer r and q. For overdamped motion, $r, q > 0$, the flux is directed down the ratchet (positive current). The ratchet effect in a noiseless ratchet ($\xi = 0$ in (3.39)) is obtained only in the underdamped regime ($m \neq 0$ in (3.39)). In the latter case, the locked trajectories correspond to $q = r = 0$, whereas the running trajectories, both deterministic and chaotic, can occur, due to inertia, in either direction with $q = \pm 1, \pm 2, \ldots$ in Eq. (3.41). The direction of the flux depends sensitively on the control parameters in Eq. (3.39).

The ratchet effect means the transformation of the fluctuation environment into deterministic directional motion. In the case considered above, deterministically induced chaos mimics the role of noise. This result is supported by an analysis [121] of the values of the control parameters for which current reversal occurs. The origin of current reversal is a bifurcation from a chaotic to a periodic regime. The same analysis has recently been extended [122] to the case in which an additional constant force is included in Eq. (3.39).

3.4.4 *Absolute negative mobility*

In the case of external constant, periodic and random forces, the equation of motion is

$$\frac{d^2\phi}{dt^2} + \alpha\frac{d\phi}{dt} + \sin\phi = a + A\sin(\Omega t) + \xi(t) \qquad (3.42)$$

When an external static force is applied to a system, it produces a current in the direction of the applied force. However, this trivial statement is correct only for a state initially in equilibrium for which the response is always in the same direction, such that the system after distortion tends to a new equilibrium. For systems far from equilibrium, such as those described by Eq. (3.42), an apparently paradoxical situation may occur [123; 124]. The system response may be directed in the direction opposite to that of a small external force (absolute negative mobility (ANM)). This effect differs from the ratchet effect which occurs for a non-symmetric potential whereas ANM occurs for the symmetric potential $\sin\phi$ in Eq. (3.42).

The running solutions of Eq. (3.42) contain both periodic and chaotic motions. Adding fluctuations leads to random transitions between possibly coexisting basins of attraction, similar to random transitions between potential wells in equilibrium systems. Numerical calculations have been performed for the asymptotic mean angular velocity

$$\langle\langle v \rangle\rangle \equiv \frac{T}{L}\left\langle \lim_{t\to\infty}\frac{1}{t}\int\limits_0^t \phi(\tau)\,d\tau \right\rangle \qquad (3.43)$$

averaged over time and thermal fluctuations, where T and L are the temporal and spatial periods of the system. ANM means that v is directed opposite to the bias force a.

For values of the parameter a lying in the interval $(-a_{cr}, a_{cr})$, the velocity is oriented in the opposite direction from that of the driving force a, thereby displaying ANM. The value of a_{cr} depends on the other parameters entering Eq. (2.24).

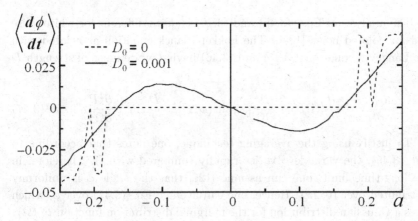

Fig. 3.3   Average angular velocity as a function of bias torque for deterministic (dashed line) and chaotic (solid line) dynamics.

The results of numerical simulations [123] are shown in Fig. 3.3. for the following values of the parameters: $\alpha = 0.9$, $A = 4.2$, $\Omega = 4.9$, without noise $(D_0 = 0)$ and for noise of strength $D_0 = 0.001$. Here, $a_{cr} = \pm 0.17$. For a fixed value of parameter a, ANM exists in some intermediate region for the strength of noise D_0 (temperature for the thermal noise) disappearing for both small and large D_0. Interestingly, the occurrence of ANM requires all the terms in Eq. (3.42). In the absence of any of these terms, ANM will not occur [124].

3.5   Pendulum with multiplicative noise

If the rod of a pendulum performs random vibrations, one has to add an external multiplicative noise $\xi(t)$ to the equation of motion,

$$\frac{d^2\phi}{dt^2} + [\omega_0^2 + \xi(t)]\sin\phi = 0 \qquad (3.44)$$

Equation (3.44) can be rewritten as two first-order stochastic differential equations

$$\frac{d\phi}{dt} = \Omega \tag{3.45}$$

$$\frac{d\Omega}{dt} = -\left[\omega_0^2 + \xi(t)\right]\sin\phi \tag{3.46}$$

The analysis of Eqs. (3.45) and (3.46) is quite different for white noise and for colored noise [125]. The Fokker-Planck equation associated with the Langevin equations (3.45) and (3.46) having white noise of strength D is

$$\frac{\partial P}{\partial t} = -\frac{\partial}{\partial\phi}(\Omega P) + \frac{\partial}{\partial\Omega}\left[\omega_0^2\sin(\phi)P\right] + \frac{D}{2}\sin^2\phi\frac{\partial^2 P}{\partial\Omega^2} \tag{3.47}$$

To justify using the averaging technique, one notes that according to Eq. (3.45), the variable ϕ varies rapidly compared with Ω. Therefore, in the long-time limit, one can assume [126] that the angle ϕ is uniformly distributed over $(0, 2\pi)$. Hence, one can average Eq. (3.47) over ϕ, which gives a Gaussian distribution for the marginal distribution function $P_1(\Omega)$,

$$P_1(\Omega) = \frac{1}{\sqrt{\pi Dt}}\exp\left(-\frac{\Omega^2}{Dt}\right) \tag{3.48}$$

Accordingly one obtains for the energy $E \approx \Omega^2$, which gives, using (3.48),

$$P_1(E) = \sqrt{\frac{2}{\pi DEt}}\exp\left(-\frac{2E}{Dt}\right) \tag{3.49}$$

From (3.49), it follows that

$$\langle E\rangle = \frac{D}{4}t \tag{3.50}$$

For colored noise, one cannot write the exact Fokker-Planck equation. However, if one assumes that ϕ has a power-law dependence as a function of time t, a self-consistent estimate gives [125],

$$\langle E\rangle \approx \sqrt{t} \tag{3.51}$$

These heuristic arguments have been confirmed [125] by more rigorous analysis and by numerical calculations.

3.6 Parametric periodic and random forces

Here we consider the influence of noise on the motion of a pendulum subject to the multiplicative periodic force discussed in Section 2.4. One can introduce noise into Eq. (2.18), both additively and multiplicatively. In the former case, Eq. (2.18) takes the following form,

$$\frac{d^2\phi}{dt^2} + \alpha\frac{d\phi}{dt} + [1 + A\sin(\Omega t)]\sin\phi = \xi(t) \qquad (3.52)$$

The random noise $\xi(t)$ has been chosen [127] to be $\xi = \xi_0\eta(t)$, where ξ_0 is the amplitude of the noise and $\eta(t)$ is a number chosen randomly at time t from a Gaussian distribution with zero mean. It is convenient to describe the influence of noise on the noise-free solution for parameters $\alpha = 0.046$, $\Omega = 2.10$, and $A = 8.95$ [128]. There are three coexisting stable periodic attractors. One is a symmetric oscillation around the downward position, and the other two are large amplitude asymmetric oscillations around the upward direction. Without noise, the pendulum can remain in any of these periodic modes indefinitely. In the presence of noise, the system hops endlessly among the periodic attractors, i.e., the system exhibits intermittent chaos driven by noise. However, external noise does not influence the chaotic transient, i.e., the characteristic time for the transition from chaotic to periodic trajectories with increasing control parameter which is determined by the random phase Ψ [129],

$$\frac{d^2\phi}{dt^2} + \alpha\frac{d\phi}{dt} + [1 + A\sin(\Omega t + \Psi)]\sin\phi = 0 \qquad (3.53)$$

For a large interval of increasing noise level, the average rotational velocity is stable: the noise component creates oscillations around the rotational average velocity [129]. However, for a large enough noise level, the rotational motion vanishes as an independent mode. In addition, one obtains [129] an intermittent transition to chaotic motion induced by noise.

Another effect resulting from noise is the conversion of the initial chaotic motion into regular motion, which terminates at one of the fixed points. Calculations have been performed [130] for high-frequency stochastic oscillations for the following parameters : $A = 2$, $\alpha = 1$, $\Omega = 0.5$, for two sets of initial conditions: $\phi_{t=0} = 0$, $(d\phi/dt)_{t=0} = 1$, and $\phi_{t=0} = 2$, $(d\phi/dt)_{t=0} = 0$. In the former case, there are nonregular librations without noise, whereas adding noise terminates the motion at the fixed point $\phi = 0$. In the latter case, nonregular librations and rotations are terminated at the fixed point

$\phi = -2\pi$. If a system originally has a stable limit cycle instead of fixed points, then the addition of noise will transform the chaotic behavior into a regular limit cycle. Analogous numerical calculations (with $\sin\phi$ replaced by $b\phi + c\phi^3$) have been performed [131] for low-frequency stochastic oscillations for the parameters: $A = 0.94$, $\alpha = 0.15$, $\Omega = 0.5$, $b = c = 1$ and initial conditions $\phi_{t=0} = 1$, $(d\phi/dt)_{t=0} = 1$. The results were similar to those obtained without noise [130]: the motion is chaotic with randomly alternating librations and rotations, whereas adding sufficiently large random noise stabilizes the system by eliminating chaos. The latter result has been confirmed [131] by the calculation of the Lyapunov exponents.

It is remarkable that in contrast to additive noise, multiplicative noise is able to convert the chaotic trajectories, induced by the deterministic chaos, into regular trajectories.

3.7 Damped pendulum subject to a constant torque, periodic force and noise

The addition of a constant term a to the right hand side of Eq. (3.34) leads to the loss of periodicity. As a increases, the system alternatively becomes chaotic and periodic. One of the two intermittent regions described in the previous section is destroyed even by a small bias a ($a = 0.0005$ for $\Omega = 0.47315$, $\alpha = 0.2$, and $A = 1.2$), and the motion becomes nearly ballistic, in contrast to the case without bias, for which two symmetric intermittent regions lead to normal diffusion [112]. These counterintuitive results have been obtained also [123] for the general case, for which, in addition to a constant term a, thermal noise is also included in Eq. (3.26), which then reverts to the original Eq. (3.42). Numerical calculations have been performed [123] for the asymptotic mean velocity $\langle\langle d\phi/dt\rangle\rangle$, which is defined as the average of the angular velocity over time and thermal fluctuations. For values of parameter a lying in the interval $(-a_{cr}, a_{cr})$, the velocity is oriented in the direction opposite to the driving force, thereby displaying "absolute negative mobility" (see Section 3.4.4). The value of a_{cr} depends on the other parameters in Eq. (3.42).

A pendulum subject to both additive noise and multiplicative noise has recently been considered [132]. The equation of motion is obtained by adding multiplicative noise to Eq. (3.42),

$$\frac{d^2\phi}{dx^2} + \gamma\frac{d\phi}{dt} + [a + \sigma\xi(t)]\sin\phi = a + f\sin(\Omega t) + \eta(t) \tag{3.54}$$

where $\xi(t)$ and $\eta(t)$ are assumed to be delta-function correlated Orenstein-Uhlenbeck sources of noise,

$$\langle \xi(t)\xi(t_1)\rangle = E^2 \exp(-\lambda|t-t_1|); \quad \langle \eta(t)\eta(t_1)\rangle = sE^2 \exp(-\lambda|t-t_1|)$$
$$\langle \xi(t)\eta(t_1)\rangle = 2D\delta(t-t_1) \tag{3.55}$$

Numerical calculations show [132] that if the product $D\lambda$ is constant, the system remains in either a locked state or a running state as λ changes. For small λ, for $t \to \infty$, the system is randomly located in one of two stable states, $\phi = 0$ or $\phi = -\pi$. For large λ, the system is always in the state $\phi = 0$, and, as λ increases, ϕ undergoes the monostability-bistability-monostability transition. The coefficient σ of multiplicative noise in (3.54) simulates the time evolution of ϕ: the running solutions start from $|\sigma| > 0$, and ϕ turns over counterclockwise at $\sigma = -20$. Furthermore, the larger the value of σ, the faster ϕ turns over. This means that the strength of multiplicative noise controls the rotation direction of ϕ without any external torque. This bistability regime exists for any value of additive noise, whose strength serves only to change the turnover direction and the speed of ϕ.

3.8 Overdamped pendulum

If the damping is strong enough, one can neglect the inertial term in the pendulum equation and consider the dimensionless dynamic equation of the overdamped pendulum of the form

$$\frac{d\phi}{dt} = a - \sin\phi + \xi(t) \tag{3.56}$$

Note that this equation has a chaotic solution due to the random force $\xi(t)$, whereas in the case of deterministic chaos, considered in the previous chapter, the overdamped pendulum does not have enough degrees of freedom to exhibit chaotic behavior.

Two quantities characterize the dynamics of noisy systems: the average velocity

$$\left\langle \frac{d\phi}{dt}\right\rangle \equiv \lim_{t\to\infty} \frac{\phi(t)}{t} \tag{3.57}$$

and the effective diffusion coefficient

$$D_{eff} \equiv \lim_{t\to\infty} \frac{1}{2t}\left\{\left\langle [\phi(t)-\langle\phi(t)\rangle]^2\right\rangle\right\} \tag{3.58}$$

One should emphasize the role of noise in Eq. (3.56). The motion of the pendulum takes place subject to a washboard potential $U(\phi) = -a\phi - \cos\phi$, consisting of wells and barriers. For small a, the motion of the particle is essentially confined to one potential well, which is equivalent to pendulum motion around the downward position. However, for large a, the particle can overcome the barrier following the driving force a (the rotation motion of a pendulum). There is a clear threshold ($a \gtrsim 1$) between these "locked" and "running" states. In the presence of noise, this threshold is blurred since even for small a, the particle is able to overcome the potential barrier to follow the driving force a. In this section we will discuss many other effects due to different types of noise [2].

3.8.1 *Additive white noise*

One can find the exact analytic solution for both $\langle d\phi/dt \rangle$ and D for the more general equation of the form

$$\frac{d\phi}{dt} = a - \frac{dU}{d\phi} + \xi(t) \tag{3.59}$$

where $U(x)$ is a periodic function with period L,

$$U(\phi + L) = U(\phi) \tag{3.60}$$

and $\xi(t)$ is white noise.

The general expression for the average velocity was obtained long ago [133],

$$\left\langle \frac{d\phi}{dt} \right\rangle = \frac{1 - \exp(aL/T)}{\int_{\phi_0}^{\phi_0 + L} I_\pm(x)(dx/L)} \tag{3.61}$$

with

$$I_+(x) = D_0^{-1} \exp\left[(U(x) - a\right] \int_{x-L}^{x} dy \exp\left\{-[(U(y) - a]\right\} \tag{3.62}$$

and

$$I_-(x) = D_0^{-1} \exp\left\{-[(U(x) - a]\right\} \int_{x}^{x+L} dy \exp\left[(U(y) - a\right] \tag{3.63}$$

where D_0 is the diffusion coefficient for Eq. (3.59) without the periodic term, and $I_\pm(x)$ means that the index may be either plus or minus.

The diffusion coefficient (3.58) that corresponds to Eq. (3.59) was obtained recently [134] by using the moments of the first passage time,

$$D = D_0 \frac{\int_{\phi_0}^{\phi_0+L} I_\pm (y) I_+ (y) I_- (y) (dy/\dot{L})}{\left[\int_{\phi_0}^{\phi_0+L} I_\pm (y) (dy/L)\right]^3} \tag{3.64}$$

The ratio of the two integrals in (3.64) can be very large [134], so that the presence of the periodic potential in Eq. (3.59) can result in an increase of the diffusion coefficient D by 14 orders of magnitude!

This effect is not only very large but also has the opposite sign compared with the analogous effect for equilibrium processes. In the latter case, the diffusion coefficient decreases upon the addition of a periodic potential acting on the Brownian particle due to localization of the particle in the periodic potential [135].

Let us return to the sinusoidal form of the torque. The influence of thermal (additive) noise on the pendulum is described by the following equation,

$$\frac{d\phi}{dt} = a_0 - b_0 \sin \phi + \xi (t). \tag{3.65}$$

The exact solution of this equation for white noise of strength D is well known [133],

$$\langle \phi \rangle = \frac{\sinh (\pi a_0/D)}{\pi/D} \left| I_{\pi a_0/D} \left(\frac{b_0}{D}\right)\right|^{-2} \tag{3.66}$$

where $I_{\pi a_0/D}$ is the modified Bessel function of first order [136]. In the limit $a_0, D << 1$, Eq. (3.61) reduces to

$$\left\langle \frac{d\phi}{dt} \right\rangle = 2 \sinh \frac{\pi a_0}{D} \exp \left(-\frac{2b_0}{D}\right). \tag{3.67}$$

Each of the two factors in Eq. (3.67) has a clear physical meaning [137]. The Arrhenius exponential rate, $\exp\left[-(2b_0/D)\right]$, decreases with decreasing D, which makes it easier for the system to overcome a potential barrier, while the pre-exponential factor — the difference between approach to the left well and to the right well — makes the system more homogeneous.

In addition to the sinusoidal form of the periodic function $U(\phi)$ in Eqs. (3.61) and (3.64), an analysis was also performed [138] for the sawtooth potential. As was the case for the periodic potential, the diffusion coefficient

D increases as a function of the tilting force a. Hence, there are two sources for an increase in diffusion, the effect of the tilting force a ("passive channel" in the terminology of [138]) and a huge enhancement coming from the periodic force ("active channel"). An additional quantity that was studied [138] is the factor of randomness $Q = 2D/(L \langle d\phi/dt \rangle)$, which defines the relation between the diffusive and directed components in the Brownian motion.

3.8.2 *Additive dichotomous noise*

For a pendulum subject to a pure periodic torque ($a = 0$ in Eq. (3.65)), one obtains

$$\frac{d\phi}{dt} + b_0 \sin \phi = \xi(t) \tag{3.68}$$

For white noise $\xi(t)$, the average flux $\langle d\phi/dt \rangle$ vanishes. Here we consider asymmetric dichotomous noise $\xi(t)$ which randomly jumps back and forth from $\xi = A$ to $\xi = -B$, with transition probabilities γ_1 and γ_2. The asymmetry can be described by the parameter ε [139],

$$A = \sqrt{\frac{D}{\tau}\left(\frac{1+\varepsilon}{1-\varepsilon}\right)}; \quad B = \sqrt{\frac{D}{\tau}\left(\frac{1-\varepsilon}{1+\varepsilon}\right)}; \quad \gamma_{1,2} = \frac{1 \pm \varepsilon}{2} \tag{3.69}$$

Equations (3.69) satisfy the condition $\gamma_1 B = \gamma_2 A$. Therefore, the requirement $\langle \xi(t) \rangle = 0$ is obeyed.

In the presence of dichotomous noise, it is convenient to define two probability densities, $P_+(\phi, t)$ and $P_-(\phi, t)$, which correspond to the evolution of $\phi(t)$ subject to noise of strength A and $-B$, respectively. The set of Fokker-Planck equations satisfied by these two functions is a slight generalization of that considered in Section 3.4.2,

$$\frac{\partial P_+}{\partial t} = -\frac{\partial}{\partial \phi}\left[(b_0 \sin \phi + A) P_+\right] - \gamma_1 P_+ + \gamma_2 P_- \tag{3.70}$$

and

$$\frac{\partial P_-}{\partial t} = -\frac{\partial}{\partial \phi}\left[(b_0 \sin \phi - B) P_-\right] - \gamma_2 P_- + \gamma_1 P_+ \tag{3.71}$$

Introducing the probability function $P = P_+ + P_-$, which satisfies the normalization and periodicity conditions, one obtains a cumbersome expression for the average angular velocity [139] which, to first order in the parameter $\sqrt{\tau/D}$, takes the following form

$$\left\langle \frac{d\phi}{dt} \right\rangle = \frac{\varepsilon}{\sqrt{1-\varepsilon^2}} \frac{\sqrt{\tau/D}}{I_0^2 (b_0/D)} \tag{3.72}$$

where I_0 is a modified Bessel function.

Another limiting case [140] where the solution has a simple form is for slow jumps, $\gamma_{1,2} \to 0$ ("adiabatic approximation"),

$$\left\langle \frac{d\phi}{dt} \right\rangle \sim \begin{cases} (\gamma_1^{-1} + \gamma_2^{-1})^{-1} \left[\dfrac{1}{\gamma_2} \sqrt{(a_0 - B) - b^2} + \dfrac{1}{\gamma_1} \sqrt{(a_0 + A)^2 - b^2} \right], \\ \qquad\qquad\qquad\qquad\qquad\qquad a_0 - B < b, \ a_0 + A < b; \\[2mm] \gamma_1 (\gamma_1^{-1} + \gamma_2^{-1})^{-1} \left(\sqrt{(a_0 + A)^2 - b^2} \right), \ a_0 - B < b, \ a_0 + A > b \end{cases} \tag{3.73}$$

These equations imply that in the adiabatic approximation, the total mobility is the average of the mobilities for the two corresponding potentials [141].

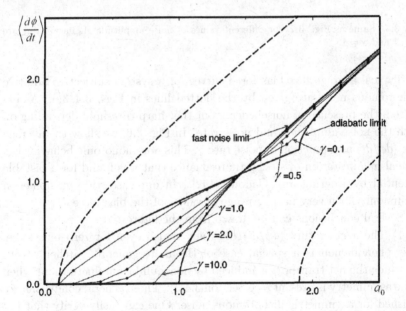

Fig. 3.4 Dimensionless average angular velocity as a function of bias with additive dichotomous noise for different values of noise rate $\gamma = \gamma_1 + \gamma_2$. Parameters are $b_0 = 1$ and $A = B = 0.9$.

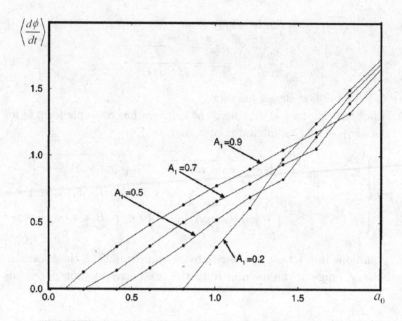

Fig. 3.5   Same as Fig. 3.2, for different values of noise amplitude A. Parameters are $b_0 = 1$ and $\gamma = 1$.

Typical average flux-bias force curves for a system subject to additive dichotomous noise are given by the dotted lines in Figs. 3.4–3.6. As expected, the presence of noise smears out the sharp threshold depending on both the noise amplitude and noise rate. In Fig. 3.6, we show the average flux $\langle d\phi/dt \rangle$ as a function of noise rate γ. This non-monotonic behavior is a special manifestation of stochastic resonance that was found for a bistable potential by Doering and Gadoua [142]. In our case, this phenomenon occurs only for a very narrow range of values of the bias force a.

Several conclusions can be drawn from the figures:

1) The average flux $\langle d\phi/dt \rangle$ does not vanish even for zero bias force. This phenomenon is a special case of the more general "ratchet effect" for which the net transport is induced by nonequilibrium fluctuations when some asymmetry is present (see Section 3.4.3). These general conditions are satisfied for asymmetric dichotomous noise. One can easily verify that the ratchet effect disappears in the limiting case of symmetric noise when $a = 0$, which leads to the vanishing of $\langle d\phi/dt \rangle$. The latter occurs both for white noise and for symmetric dichotomous noise, since the function $\Gamma(\phi)$ defined in (3.79) is an odd function for $a = 0$, which implies that $\langle d\phi/dt \rangle = 0$.

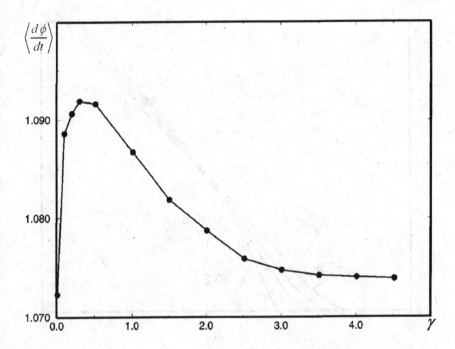

Fig. 3.6   Dimensionless average angular velocity as a function of noise rate γ for $b_0 = 1$ and $A = B = 0.9$.

The ratchet effect might have practical applications for superconducting electronics, as well as in other fields of physics, chemistry and biology (for some recent references, see [119]).

2) A stochastic resonance phenomenon (non-monotonic behavior of the average flux as a function of noise rate) has been found in a narrow region of noise rate, as shown in Fig. 3.6.

3.8.3   *Multiplicative dichotomous noise*

For a pendulum subject to multiplicative noise of the form (3.69), one obtains

$$\frac{d\phi}{dt} = a_0 - \xi(t) \sin \phi \tag{3.74}$$

Typical graphs of flux-bias characteristics for multiplicative dichotomous noise are shown in Figs. 3.7–3.9. The first two figures show $\langle d\phi/dt \rangle$

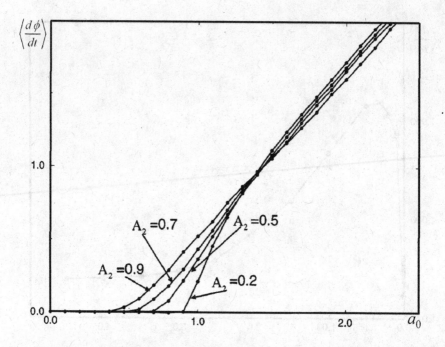

Fig. 3.7   Dimensionless average angular velocity as a function of bias with multiplicative dichotomous noise for different noise amplitudes.

as a function of a_0, which is characteristic of multiplicative dichotomous noise for different amplitudes and different noise rates, respectively. The non-monotonic behavior of $\langle d\phi/dt \rangle$ as a function of noise rate γ is shown in Fig. 3.9. Just as for additive noise, stochastic resonance occurs in a narrow regime of the noise rate γ.

Note the non-trivial type of ratchets (discussed in Section 3.4.3) exists in this case. In the absence of additive noise but in presence of symmetric multiplicative noise, $f_2(t) = \pm b$, our basic equation is

$$\frac{d\phi}{dt} = a_0 - f_2(t)\sin\phi \qquad (3.75)$$

If $a_0 < |b|$, Eq. (3.75) gives $\langle d\phi/dt \rangle = 0$ for both $f_2(t) = \pm b$. However, if one allows switching between two dynamics situations, the resulting motion will have a net average angular velocity. This can be seen from Fig. 3.10, where the two washboard potentials $V_\pm = a\phi \pm b\cos\phi$ are shown. If, as usually assumed, the rate of reaching the minimal energy in each well is much larger than γ (adiabatic approximation), $\langle d\phi/dt \rangle$ is non-zero for

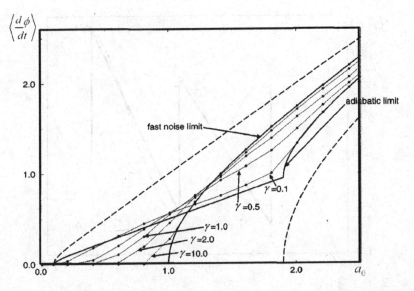

Fig. 3.8 Same as Fig. 3.5, for different values of noise rate γ for $b_0 = 1$ and $A = B = 0.9$.

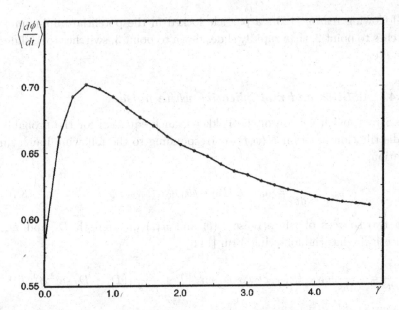

Fig. 3.9 Dimensionless average angular velocity as a function of noise rate for multiplicative dichotomous noise for $b_0 = 1$ and $A = B = 0.9$.

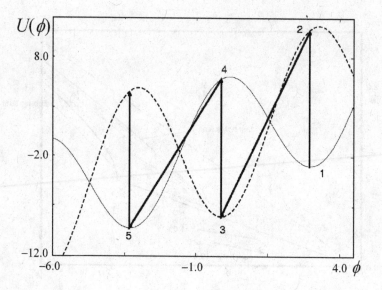

Fig. 3.10   Washboard potential $U(\phi) = a\phi - b\cos\phi$ with $a = 1$ and $b = 7$. A particle cannot move along each of the potentials by itself, but if one allows a transition between potentials, the particle moves downhill along the trajectory $1 : 2 : 3 : 4 : 5$.

the following reason [143]: a particle locked in the potential minimum 1, switches to point 2, then rapidly slides down to point 3, switches to 4, slides to 5, etc.

3.8.4   *Additive and multiplicative white noise*

The Stratonovich version of the Fokker-Planck equation for the probability distribution function $P(\phi, t)$, corresponding to the following Langevin equation,

$$\frac{d\phi}{dt} = [a_0 + \xi(T)] - [b_0 + \eta(t)]\sin\phi \qquad (3.76)$$

with two sources of white noise $\xi(t)$ and $\eta(t)$ of strength D_1 and D_2, respectively, has the following form [144],

$$\frac{\partial P}{\partial t} = -\frac{\partial}{\partial\phi}[a_0 - (b_0 + D_2\cos\phi)\sin\phi]P + \frac{\partial^2}{\partial\phi^2}\left[(D_1 + D_2\sin^2\phi)\right]P$$

$$\equiv -\frac{\partial J}{\partial\phi}, \qquad (3.77)$$

where J is the flux proportional to $\langle d\phi/dt \rangle$, that is, $\langle d\phi/dt \rangle = 2\pi J$.

For the stationary case, $\partial P/\partial t = 0$, the distribution function satisfies the following differential equation [140],

$$\frac{dP_{st}}{dx} + \Gamma(\phi) P_{st} = J\Omega^2(\phi), \qquad (3.78)$$

where

$$\Gamma(\phi) = -\frac{a_0 - b_0 \sin\phi - D_2 \sin\phi \cos\phi}{(D_1 + D_2 \sin^2\phi)}; \qquad \Omega(\phi) = (D_1 + D_2 \sin^2\phi)^{-1/2} \qquad (3.79)$$

The solution of the first-order differential equation (3.78) contains one constant which, together with the constant J, are determined from the normalization condition, $\int_{-\pi}^{\pi} P(\phi)\, d\phi = 1, P(-\pi) = P(\pi)$. The expression for P_{st} can be accurately approximated for small noise by the method of steepest descent [145],

$$P_{st} \simeq C\left[\left(1 + \frac{D_2}{D_1}\sin^2\phi\right)^{\frac{1}{2}} + (a_0 + b_0\sin\phi)/D_1\right]^{-1} \qquad (3.80)$$

where C is the normalization constant. For $D_2 < D_1$, the term $a_0 + b_0\sin\phi$ makes the main contribution to P_{st}. For $|a_0| = |b_0|$, P_{st} contains a single maximum. For $D_2 > D_1$, the main contribution to P_{st} is the term $[1 + (D_2/D_1)\sin^2\phi]^{1/2}$, i.e., P_{st} has maxima at the points $n\pi$ for integer n. Additive and multiplicative noise have opposite influences on P_{st} [145]. An increase of multiplicative noise leads to the increase and narrowing of the peaks of P_{st}, whereas the increase of additive noise leads to their decrease and broadening.

Calculations yield [140]

$$\left\langle \frac{d\phi}{dt} \right\rangle = \frac{2\pi\left[1 - \exp\left(-2\pi a_0/\sqrt{D_1(D_1 + D_2)}\right)\right]}{\left[\int_0^{2\pi} \Omega(x) F(x,0) \left(\int_x^{x+2\pi} \Omega(y) F(0,y)\, dy\right) dx\right]} \qquad (3.81)$$

where

$$F(k,l) = \exp\left[-\int_k^l T(z)\, dz\right]; \qquad T(z) = \frac{a_0 - b_0\sin z}{D_1 + D_2\sin^2 z} \qquad (3.82)$$

We have performed numerical calculations of $\langle d\phi/dt \rangle$ in order to compare the importance of additive and multiplicative noise [146]. The average

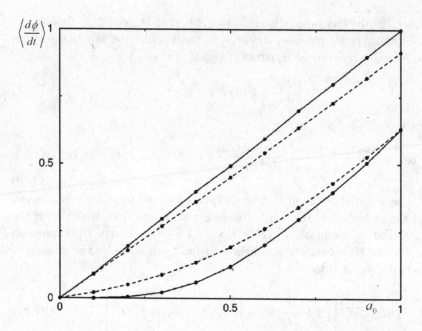

Fig. 3.11 Average angular velocity as a function of bias for $b_0 = 1$. The solid and dotted lines describe single multiplicative noise and single additive noise, respectively. The upper and lower curves correspond to strengths 2 and 0.1, respectively.

angular velocity $\langle d\phi/dt \rangle$ and a_0 correspond to the voltage and the bias current for a Josephson junction. Therefore, the graph $\langle d\phi/dt \rangle$ versus a_0 gives the voltage-current characteristic of a junction. In Fig. 3.11, we display $\langle d\phi/dt \rangle$ as a function of a_0 for $b_0 = 1$, in the presence of one of the two sources of noise given by $D = 0.1$ and $D = 2.0$. One sees that for a small value of noise ($D = 0.1$), additive noise leads to higher flux than multiplicative noise, whereas for a larger value of noise ($D = 2$), the opposite occurs. The transition takes place for an intermediate value of noise. As shown in Fig. 3.12, for noise of strength $D = 1$ (of order b_0), additive noise produces a larger average angular velocity for small a_0 and the opposite effect for large a_0.

It is convenient to consider separately the two limiting cases of weak ($D_1 \to 0$) and strong ($D_1 \to \infty$) additive noise, combining analytic and numerical calculations. Let us start with the case of weak noise, $D_1 \to 0$ and $D_2 \to 0$, with either $D_2 > D_1$ or $D_2 < D_1$. Calculating the integrals

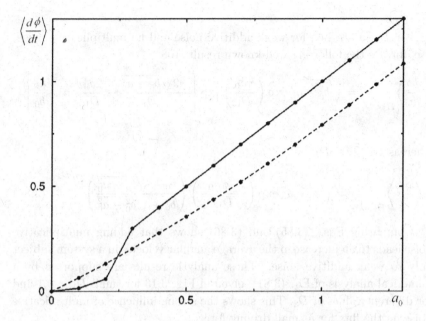

Fig. 3.12   Same as Fig. 3.9 in the presence of a single source of noise of strength 1.

in Eq. (3.81) by the method of steepest descent yields

$$\left\langle \frac{d\phi}{dt} \right\rangle = \left[1 - \exp\left(-\frac{2\pi a_0}{\sqrt{D_1\left(D_1 + D_2\right)}} \right) \right] \frac{\sqrt{\left| T\left(z_{\max}\right) T\left(z_{\min}\right) \right|}}{\Omega\left(z_{\max}\right) \Omega\left(z_{\min}\right)}$$

$$\times \exp \int_{z_{\max}}^{z_{\min}} T\left(z\right) dz \tag{3.83}$$

where $z_{\min}$ and $z_{\max}$ are two neighboring zeros of $T\left(z\right)$ with $T\left(z_{\max}\right) > 0$, $T\left(z_{\min}\right) < 0$.

It is easily found from Eq. (3.82) that $\sin\left(z_{\max,\min}\right) = a_0/b_0$, $\cos z_{\max} = \omega/b_0$, $\cos z_{\min} = -\omega/b_0$. For $b_0 > a_0$, Eq. (3.83) reduces to

$$\left\langle \frac{d\phi}{dt} \right\rangle = \sqrt{b_0^2 - a_0^2} \left[1 - \exp\left(-\frac{2\pi a_0}{\sqrt{D_1\left(D_1 + D_2\right)}} \right) \right] \exp \int_{z_{\max}}^{z_{\min}} T\left(z\right) dz$$

$$\tag{3.84}$$

One can evaluate the integral in (3.84). However, instead of writing out this cumbersome expression, we present the results for the two limiting cases of large and small multiplicative noise compared with additive noise,

$D_2 \lesssim D_1$.

For $D_2 < D_1$, i.e., for weak additive noise and no multiplicative noise, one obtains the following well-known result [108],

$$\left\langle \frac{d\phi}{dt} \right\rangle_{D_2 < D_1} = \sqrt{b_0^2 - a_0^2} \exp\left(\frac{\pi a_0}{D_1}\right) \exp\left[-\frac{2\sqrt{b_0^2 - a_0^2}}{D_1} - \frac{2a_0}{D_1} \sin^{-1} \frac{a_0}{b_0}\right]$$

$$(3.85)$$

whereas for $D_2 > D_1$,

$$\left\langle \frac{d\phi}{dt} \right\rangle_{D_2 > D_1} = \sqrt{b_0^2 - a_0^2} \exp\left(\frac{-\pi a_0}{\sqrt{D_1 D_2}}\right) \left(\frac{b_0 - \sqrt{b_0^2 - a_0^2}}{b_0 + \sqrt{b_0^2 - a_0^2}}\right)^{b_0/D_2} \quad (3.86)$$

Comparing Eqs. (3.85) and (3.86) shows that adding multiplicative noise leads to an increase in the average angular velocity in a system subject only to weak additive noise. These analytic results are supported by a numerical analysis of Eq. (3.81), given in Fig. 3.13 for small $D_1 = 0.1$ and for different values of D_2. This shows the strong influence of multiplicative noise on the flux for a small driving force.

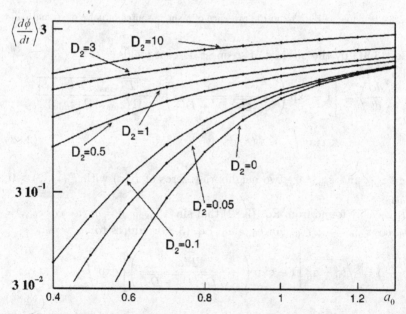

Fig. 3.13 Dimensionless average angular velocity as a function of bias for $b_0 = 1$, $D_1 = 0.1$ for different values of D_2.

Turning now to the opposite limiting case of strong additive noise, $D_1 \to \infty$, one can greatly simplify Eq. (3.81),

$$\left\langle \frac{d\phi}{dt} \right\rangle_{D_1 \to \infty} = \frac{a_0 \pi^2}{D_1} \left(1 + \frac{D_2}{D_1} \right)^{-\frac{1}{2}} \left[\int_0^\pi \Omega(z)\, dz \right]^{-2} \qquad (3.87)$$

Figure 3.14 displays the dimensionless average angular velocity $(1/a_0) \langle d\phi/dt \rangle$ as a function of D_2/D_1 for large additive noise D_1. The curve starts from $(1/a_0) \langle d\phi/dt \rangle_{D_1 \to \infty} = 1$ for $D_2 = 0$ (large additive noise suppresses the sin term in Eq. (2.6), yielding Ohm's law for the Josephson junction [108]), and increases markedly as the strength of multiplicative noise increases.

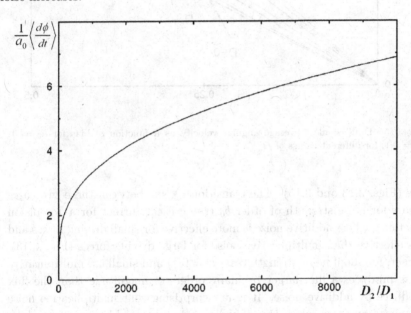

Fig. 3.14 Dimensionless average angular velocity as a function of the ratio of noise strengths D_2/D_1.

Figure 3.15 displays the results of the numerical analysis of Eq. (3.81) for comparable values of all parameters (b_0, D_1 and D_2), which again demonstrates an increase of the flux due to multiplicative noise.

One concludes that in the presence of one source of noise, the average angular velocity $\langle d\phi/dt \rangle$ is larger for additive noise if the strength of the noise is small, whereas for strong noise, multiplicative noise is more effec-

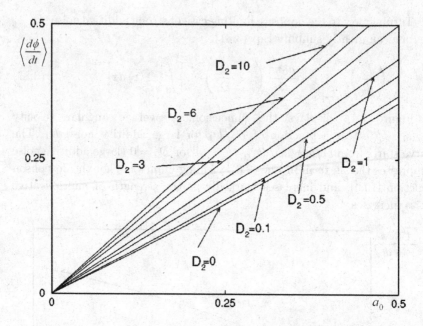

Fig. 3.15   Dimensionless average angular velocity as a function of bias for $a_0 = 1$, $D_1 = 0.1$, for different values of D_2.

tive (Figs. 3.13 and 3.15). The transition regime between these two cases occurs for noise strength of order b_0 (the critical current for a Josephson junction), where additive noise is more effective for small driving forces and less effective than multiplicative noise for large driving forces (Fig. 3.13). In fact, for small noise strength (say, $D = 0.1$) and small a_0, multiplicative noise produces a flux larger by many orders of magnitude than the flux produces by additive noise. It is not surprising that multiplicative noise becomes important when D is of order of the potential barrier height b_0.

If both sources of noise are present, then the flux is increased by the presence of strong multiplicative noise for weak (Fig. 3.13), strong (Fig. 3.14) and intermediate (Fig. 3.15) strength of additive noise, especially for small bias force a_0. The latter result has a simple intuitive explanation. Indeed, the horizontal periodic potential ($a_0 = 0$) with strong fluctuations in the width of this potential has no preferred direction and, therefore, the average angular velocity vanishes, $\langle d\phi/dt \rangle = 0$. It is sufficient to have a small slope of a periodic potential, $a_0 \neq 0$, for $\langle d\phi/dt \rangle \neq 0$ to occur.

The importance of multiplicative noise for stationary states has long been known [147]. The influence of both additive and multiplicative noise on the escape time from a double-well potential has also been discussed [148; 149]. The analysis of the stationary probability distribution function for a periodic potential and dichotomous multiplicative noise was given by Park et al. [150]. We have studied [140] the influence of both additive and multiplicative noise on the voltage-current characteristics of Josephson junctions. A similar effect for the output-input relation for motion in a double-well potential has been studied intensively by two groups of researchers, who called this effect "noise-induced hypersensitivity" [151] and "amplification of weak signals via on-off intermittency" [152].

3.8.5 *Multiplicative dichotomous noise and additive white noise*

There is a special case in which one or both sources of noise are dichotomous. Consider the case in which the multiplicative noise $\eta(t)$ in Eq. (3.76) is dichotomous and the noise $\xi(t)$ is white [150; 153],

$$\langle \eta(t)\eta(t+\tau)\rangle = \Delta^2 \exp(-\lambda|\tau|); \quad \langle \xi(t)\xi(t+\tau)\rangle = 2\sigma^2\delta(\tau) \quad (3.88)$$

It is convenient to define two probability densities, $P_+(\phi, t)$ and $P_-(\phi, t)$, that correspond to the evolution of $\phi(t)$ subject to noise of strength Δ and $-\Delta$, respectively. The Fokker-Planck equations for $P_+(\phi, t)$ and $P_-(\phi, t)$ are

$$\frac{\partial P_+}{\partial t} = -\frac{\partial}{\partial \phi}\left\{[a_0 - (b_0 + \Delta)\sin\phi]P_+ - \sigma^2\frac{\partial P_+}{\partial \phi}\right\} - \lambda(P_+ - P_-) \quad (3.89)$$

$$\frac{\partial P_-}{\partial t} = -\frac{\partial}{\partial \phi}\left\{[a_0 - (b_0 - \Delta)\sin\phi]P_- - \sigma^2\frac{\partial P_-}{\partial \phi}\right\} + \lambda(P_+ - P_-)$$

Equations (3.89) can be replaced by the equations for $P = P_+ + P_-$ and $Q = P_+ - P_-$,

$$\frac{\partial P}{\partial t} = -\frac{\partial}{\partial \phi}\left\{[-a_0 + b_0\sin\phi]P - \Delta Q\sin\phi - \sigma^2\frac{\partial P}{\partial \phi}\right\} \equiv -\frac{\partial J}{\partial \phi} \quad (3.90)$$

$$\frac{\partial Q}{\partial t} = -\frac{\partial}{\partial \phi}\left\{[-a_0 + b_0\sin\phi]Q - \Delta P\sin\phi - \sigma^2\frac{\partial Q}{\partial \phi}\right\} - 2\lambda Q$$

In the limit $\Delta, \lambda \to \infty$ with $\Delta^2/2\lambda \equiv D$, one recovers the results considered previously for two sources of white noise. For $a_0 = 0$, the stationary probability density $P_{st}(\phi)$ has the following form

$$P_{st}(\phi) = C \frac{1}{\sqrt{1+k^2\sin\phi}} \left[\frac{\sqrt{1+k}+\sqrt{k}\cos\phi}{\sqrt{1+k}-\sqrt{k}\cos\phi}\right]^{b_0/2D\sigma\sqrt{1+k}} \tag{3.91}$$

where $k = D^2/\sigma^2$ and C is the normalization constant. If the strength of the multiplicative noise is small, $D < D_{cr}$, $P_{st}(\phi)$ has a simple maximum at $\phi = 0$. For $D > D_{cr}$, $P_{st}(\phi)$ has double maxima at $\phi = 0$ and π, and a minimum at $\phi = \cos^{-1}\left(-b_0^2/D\right)$. The qualitative behavior of $P_{st}(\phi)$ does not depend on the strength σ^2 of the additive noise.

For both multiplicative dichotomous noise and additive white noise, Eqs. (3.89) must be solved numerically. Such calculations have been performed [150] for $\sigma = 0.1$. For given b_0 and Δ, $P_{st}(\phi)$ undergoes a phase transition from the double-peak state to the single-peak state upon increasing the correlation time λ. The physical explanation of such behavior is as follows: for fast processes, the system is under the influence of the average noise and, therefore, it tends to be attracted to the fixed point $\phi = 0$ (single-peak state). For slow processes and strong noise strength, the system spends most of its time at the two fixed points $\phi = 0$ and $\phi = \pi$ (double-peak state).

A detailed analysis has been performed [140], whose main results are:

1. Two sources of dichotomous noise are able to produce a flux for small bias, whereas each by itself is unable to produce a flux in this region of bias. This effect also occurs for two sources of white noise.

2. The simultaneous action of two sources of noise can be larger than each source by itself in some region of the average flux-bias plane, but smaller in other regions.

3. The "ratchet effect" occurs for asymmetric additive noise in the presence of multiplicative noise. The latter eases the requirement for the onset of the ratchet.

3.8.6   *Correlated additive noise and multiplicative noise*

Thus far, we have considered additive noise and multiplicative noise as being independent. Correlations between different sources of noise may occur when they both have the same origin, as in laser dynamics [154], or when strong external noise leads to an appreciable change in the internal structure of the system and hence in its internal noise. For the simplest case of the two sources of white noise, $\xi(t)$ and $\eta(t)$ in Eq. (3.76) with the same type of correlations, one obtains

$$\langle \xi(t_1)\,\xi(t_2) \rangle = 2D_1\delta(t_1 - t_2); \quad \langle \eta(t_1)\,\eta(t_2) \rangle = 2D_2\delta(t_1 - t_2) \quad (3.92)$$
$$\langle \xi(t_1)\,\eta(t_2) \rangle = 2\lambda\sqrt{D_1 D_2}\,\delta(t_1 - t_2)$$

where the coefficient λ $(0 \le \lambda \le 1)$ indicates the strength of the correlation. A simple calculation [155] yields the solution for correlated multiplicative noise and additive noise in term of the results (3.77)-(3.82) for a system with non-correlated noise (see also [156]). Finally, the steady-state average angular velocity is

$$\left\langle \frac{d\phi}{dt} \right\rangle = \left\{ \int_0^{2\pi} \frac{\left[\int_{\phi_1}^{\phi_1+2\pi} \exp\left[\Psi(\phi_2)\right] d\phi_2 \right] d\phi_1}{B(\phi_1)\exp\left[\Psi(\phi_1)\right] - B(\phi_1 + 2\pi)\exp\left[\Psi(\phi_1 + 2\pi)\right]} \right\}^{-1},$$

$$(3.93)$$

where

$$\Psi(z) = -\int_0^z dx\, A(x)/B(x); \quad B(x) = D_2\sin^2 x - 2\lambda\sqrt{D_1 D_2}\sin x + D_1$$

$$A(x) = a_0 - b_0\sin x - \lambda\sqrt{D_1 D_2}\cos x + D_2\sin x\cos x \quad (3.94)$$

Extensive numerical calculations have been carried out [156] for non-correlated noise ($\lambda = 0$) which lead to the conclusions described in Section 3.8.4, and also for $-1 \le \lambda < 0$ and $0 < \lambda \le 1$, which lead to new phenomena.

1. **Reversal of** $\langle d\phi/dt \rangle$. For non-zero values of λ, the direction of $\langle d\phi/dt \rangle$ reverses when the ratio D_1/D_2 increases.

2. **Existence of extremum.** As D_1/D_2 increases, $\langle d\phi/dt \rangle$ possesses a minimum which changes from negative to positive values for $\lambda > 0$, and a maximum which changes from positive to negative values for $\lambda < 0$. Both the maximum and minimum exist for completely correlated noise, $\lambda = 1$.

3. **Symmetric dependence of** $\langle d\phi/dt \rangle$ **as a function of** D_1/D_2. Calculations of $\langle d\phi/dt \rangle$ have also been performed [157] for the case in which the multiplicative white noise $\eta(t)$ in (3.92) is replaced by dichotomous noise.

The following general comment should be made. The non-zero steady-state angular velocity is obtained from Eq. (3.68) which does not contain any driving force. According to the second law of thermodynamics, such noise-induced motion cannot appear in an equilibrium system which includes symmetric thermal noise. A non-zero average velocity exists only in the case of asymmetric thermal noise, as seen in Section 3.8.2. Moreover, even in a non-equilibrium state, a non-zero average velocity can appear only in the presence of symmetry breaking, for the following reason. If $\phi(t)$ is the solution of the dynamic equation for a given amplitude of noise, then $-\phi(t)$ is also a solution for t replaced by $-t$. These two solutions will give the same average velocities equal to $\pm \langle d\phi/dt \rangle$, which implies $\langle d\phi/dt \rangle = 0$. A non-zero value can be obtained when either the potential energy is non-symmetric ("ratchet" potential) or in the presence of non-symmetric dichotomous noise. As we have seen, another possibility is the existence of a correlation between additive and multiplicative noise. The latter has been convincingly demonstrated recently [158] for an overdamped pendulum subject to correlated sources of noise (3.92) moving in the symmetric potential of the form

$$u(\phi) = \frac{2\Delta V}{L}\phi - 2(n-1)\Delta V; \qquad (n-1)L \leq \phi \leq \left(n - \frac{1}{2}\right)L \quad (3.95)$$

$$u(\phi) = -\frac{2\Delta V}{L} + 2n\Delta V; \qquad \left(n - \frac{1}{2}\right)L \leq \phi \leq nL$$

where ΔV and L are the height and the period of a potential, and n is an integer. The results of analytical calculations similar to (3.93)-(3.94) and the numerical solution of the original Fokker-Planck equations are displayed in Fig. 3.16. One sees the dependence of the average flux on the strength of the correlation, which leads to a non-zero flux even in the absence of a constant driving force ($F = 0$).

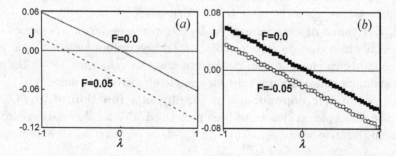

Fig. 3.16   Analytic a) and numerical b) values of the probability current as a function of the correlation length between multiplicative and additive noises for two different values of an additive static bias F (denoted a_0 in Eq. (3.76)) and for height $\Delta V = 0.125$ and period $L = 1.0$ of the potential. $D_1 = 0.1$ and $D_2 = 0.3$ are the strengths of the additive and the multiplicative noise, respectively. Reprinted from [158], Copyright (2009), with permission from Elsevier.

Chapter 4

Systems with Two Degrees of Freedom

4.1 Spring pendulum

In previous chapters we considered the chaotic behavior of a pendulum with one degree of freedom, the angle ϕ. Next in order of complexity is a spring (elastic or extensional) pendulum which is defined as a simple pendulum with a bob of mass m hanging from a spring of a stiffness constant κ inserted in its rod (Fig. 4.1).

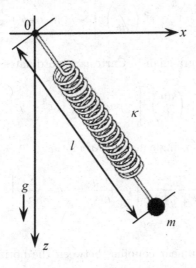

Fig. 4.1 The spring pendulum.

This system has two degrees of freedom, the coordinates x, y or the angle ϕ and the length of the rod. In the equilibrium downward position,

113

the stretched length l, due to the weight mg of the rod, is equals $l_0 + mg/\kappa$, where l_0 is the unstretched length. The well-known property of an elastic pendulum is periodic oscillations ("parametric resonance") about the upper position, unstable for a mathematical pendulum, which occurs when the spring frequency $\omega_s = \sqrt{\kappa/m}$ is about twice the pendulum frequency $\omega_0 = \sqrt{g/l}$. Chaos appears as result of the nonlinear connection between oscillating and elastic modes, and this connection is the strongest precisely in the parametric resonance region.

4.1.1   *Dynamic equations*

To analyze a spring pendulum with vertical oscillations of the suspension point, we previously [159] wrote the Lagrangian in polar coordinates,

$$L = \frac{m}{2}\left[\left(\frac{dr}{dt}\right)^2 + r^2\left(\frac{d\phi}{dt}\right)^2\right] + mgr\cos\phi - \frac{\kappa}{2}\left(r - l_0\right)^2 \qquad (4.1)$$

with energy E,

$$E = \frac{m}{2}\left[\left(\frac{dr}{dt}\right)^2 + r^2\left(\frac{d\phi}{dt}\right)^2\right] - mgr\cos\phi + \frac{\kappa}{2}\left(r - l_0\right) \qquad (4.2)$$

Here, we write the Lagrangian in Cartesian coordinates [160]

$$L = \frac{m}{2}\left[\left(\frac{dx}{dt}\right)^2 + \left(\frac{dy}{dt}\right)^2\right] - mgy - \frac{\kappa}{2}\left[\left(x^2 + y^2\right)^{1/2} - l_0\right]^2, \qquad (4.3)$$

which leads to the following equations of motion:

$$\frac{d^2x}{dt^2} = -\frac{\kappa}{m}x\left[1 - \frac{l_0}{\left(x^2 + y^2\right)^{1/2}}\right]; \quad \frac{d^2y}{dt^2} = -g - \frac{\kappa}{m}x\left[1 - \frac{l_0}{\left(x^2 + y^2\right)^{1/2}}\right]$$
$$(4.4)$$

There is a strong nonlinear coupling between the horizontal and vertical motions. For simplicity, we neglect dissipation, and therefore the energy of the system E is conserved. One obtains from (4.3),

$$E = \frac{m}{2}\left[\left(\frac{dx}{dt}\right)^2 + \left(\frac{dy}{dt}\right)^2\right] + mgy + \frac{\kappa}{2}\left[\left(x^2 + y^2\right)^{1/2} - l_0\right]^2 \qquad (4.5)$$

The energy is minimal when the spring pendulum is hanging downward with $x = 0$ and $y = -(l_0 + mg/\kappa)$,

$$E_{\min} = -mg\left(l_0 + \frac{mg}{\kappa}\right) + \frac{\kappa}{2}\left(\frac{mg}{\kappa}\right)^2 = -mg\left(l_0 + \frac{mg}{2\kappa}\right) \qquad (4.6)$$

There is no upper limit to the energy. Therefore, the characteristic parameter R, defined as

$$R = -\frac{E}{E_{\min}}, \qquad (4.7)$$

ranges from -1 to infinity.

To describe the oscillations of the bob near its downward position, it is convenient to change the coordinates $x = 0$, $y = -l$ to $x = X$ and $y = -l + Y = -l_0 - g/\omega_s^2 + Y$. Substituting these changes into (4.3), one obtains the following Lagrangian of the oscillating system,

$$L = \frac{m}{2}\left[\left(\frac{dX}{dt}\right)^2 + \left(\frac{dY}{dt}\right)^2\right] - mg\left(Y - l_0 - \frac{g}{\omega_s^2}\right) - \frac{\kappa}{2}(R_1 - l_0)^2 \qquad (4.8)$$

with $R_1 = \sqrt{X^2 + (Y + l_0 + g/\omega_s^2)^2}$. The Lagrangian equations of motion are

$$\frac{d^2X}{dt^2} = -\omega_s^2\left(1 - \frac{l_0}{R_1}\right)X \qquad (4.9)$$

$$\frac{d^2Y}{dt^2} = -\omega_s^2\left(1 - \frac{l_0}{R_1}\right)\left(Y + l_0 + \frac{g}{\omega_s^2}\right) - g \qquad (4.10)$$

The quantity $1 - l_0/R_1$, can be rewritten as

$$1 - \frac{l_0}{R_1} = 1 - \frac{l_0}{(l_0 + g/\omega_s^2 + Y)}\left[1 + \frac{X^2}{(l_0 + g/\omega_s^2 + Y)^2}\right]^{-1/2} \qquad (4.11)$$

Inserting (4.11) into (4.9) and (4.10) yields

$$\frac{d^2X}{dt^2} = -\omega_s^2 X\left\{1 - \frac{1}{(1 + \omega_0^2/\omega_s^2 + Y/l_0)}\left[1 + \frac{X^2}{l_0^2(1 + \omega_0^2/\omega_s^2 + Y/l_0)^2}\right]^{-1/2}\right\} \qquad (4.12)$$

$$d^2Y/dt^2 = -\omega_s^2 \left(Y + l_0 + g/\omega_s^2 \right) \tag{4.13}$$

$$\times \left\{ 1 - \left(1 + 1 + \omega_0^2/\omega_s^2 + Z/l_0 \right)^{-1} \left[1 + \frac{X^2}{l_0^2 \left(1 + \omega_0^2/\omega_s^2 + Y/l_0 \right)^2} \right]^{-1/2} \right\} - g$$

Thus far, all expressions are exact. For small oscillations, $X \ll l_0$, one can expand the square root of (4.12) in a power series in X,

$$\frac{d^2X}{dt^2} = -\omega_s^2 X \left[\frac{\omega_0^2/\omega_s^2 + Y/l_0}{1 + \omega_0^2/\omega_s^2 + Y/l_0} - \frac{\omega_s^6 X^2}{2l_0^2 \left(\omega_s^2 + \omega_0^2 + \omega_s^2 Y/l_0 \right)^3} \right] \tag{4.14}$$

If Y is also much smaller than l_0 $(X, Y \ll l_0)$,

$$\frac{d^2X}{dt^2} + \left[\frac{\omega_0^2 \omega_s^2}{\omega_s^2 + \omega_0^2} + \frac{\omega_s^2}{l_0} \left(\frac{\omega_s^2}{\omega_s^2 + \omega_0^2} \right)^2 Y \right] X = 0 \tag{4.15}$$

Performing the analogous expansions in Eq. (4.13), and inserting $g = \omega_0^2 l_0$,

$$\frac{d^2Y}{dt^2} + \omega_s^2 Y = -\frac{\omega_s^6 X^2}{2l_0 \left(\omega_s^2 + \omega_0^2 \right)^2} \tag{4.16}$$

Without nonlinear terms, Eqs. (4.15) and (4.16) describe the "spring" and "oscillatory" modes of a spring pendulum, whereas the nonlinear term describes the simplest form of interaction between them.

In order to keep frequencies ω_0 and ω_s independent, we have defined the oscillator frequency ω_0^2 relative to the constant parameter l_0, $\omega_0^2 = g/l_0$, in contrast to the frequency $\hat{\omega}_0^2$, which is defined relative to the variable length l, $\hat{\omega}_0^2 = g/l$, as is usually done. Using $l = l_0 \left(1 + \omega_0^2/\omega_s^2 \right)$,

$$\omega_s^2 - \hat{\omega}_0^2 = \omega_s^2 - \frac{g}{l_0 + g/\omega_s^2} = \frac{\omega_s^4}{\omega_s^2 + \omega_0^2} \tag{4.17}$$

One can rewrite Eqs. (4.15) and (4.16),

$$\frac{d^2X}{dt^2} + \hat{\omega}_0^2 X = \frac{\hat{\omega}_0^2 - \omega_s^2}{l} XY, \tag{4.18}$$

$$\frac{d^2Y}{dt^2} + \omega_s^2 Y = \frac{\hat{\omega}_0^2 - \omega_s^2}{2l} X^2, \tag{4.19}$$

which agrees with the equations of motion for the spring oscillator obtained by a slightly different method [161]. Equations (4.18) and (4.19) represent two coupled oscillators X and Y with the simplest nontrivial coupling between their Hamiltonians of the form X^2Y, which arises in other fields as well.

For the special relation $\omega_s = 2\omega_0$, the spring and the pendulum are in parametric resonance as the energy transfers back and forth from the spring mode to the oscillating mode. For this case, Eqs. (4.18) and (4.19) take the following form,

$$\frac{d^2X}{dt^2} + \hat{\omega}_0^2 X = -\left(3\hat{\omega}_0^2/l\right) XY; \qquad \frac{d^2Y}{dt^2} + 4\hat{\omega}_0^2 Y = -\left(3\hat{\omega}_0^2/2l\right) X^2 \quad (4.20)$$

Since $X, Y \ll l_0$, one can solve these equations using perturbation theory, assuming that the solution of the homogeneous equation oscillates with slowly-varying amplitude and phase [162],

$$X = A(t) \cos\left[\hat{\omega}_0 t + \psi(t)\right]; \quad Y = B(t) \cos\left[2\hat{\omega}_0 t + \chi(t)\right] \quad (4.21)$$

Inserting (4.21) into (4.20) and neglecting the small nonresonant driving force, one obtains ψ and χ, the terms linear in cos and sin, to first order in A and B. Equating the coefficients to zero gives

$$\frac{dA}{dt} = \frac{3}{4l}\hat{\omega}_0 AB \sin(2\psi - \chi); \qquad \frac{dB}{dt} = -\frac{3}{16l}\hat{\omega}_0 A^2 \sin(2 - \chi) \quad (4.22)$$

$$\frac{d\psi}{dt} = \frac{3}{4l}B \cos(2\psi - \chi); \qquad \frac{d\chi}{dt} = \frac{3A^2}{16lB}\hat{\omega}_0 \cos(2\psi - \chi) \quad (4.23)$$

Eliminating the sin terms from Eqs. (4.22) yields

$$\frac{d}{dt}\left(A^2 + 4B^2\right) = 0 \quad (4.24)$$

which expresses the conservation of energy,

$$A^2 + 4B^2 = M_0^2 \quad (4.25)$$

From Eqs. (4.23), one obtains the second constant of motion

$$A^2 B \cos(2\psi - \chi) = N_0. \quad (4.26)$$

The time dependence of the amplitude A can be found [162] from Eqs. (4.22), (4.25) and (4.26),

$$\frac{1}{2}\left(\frac{d\alpha}{d\tau}\right)^2 + V(\alpha) = E \quad (4.27)$$

where E is a constant, and

$$\alpha = \frac{A^2}{M_0^2}; \quad \tau = \frac{3\hat{\omega}_0 M_0}{4\sqrt{2l}}t; \quad V(\alpha) = -\alpha^2 + \alpha^3 \tag{4.28}$$

According to (4.28) and (4.25), $0 < \alpha < 1$. The two turning points α_0 and α_1 of the function $A = A(t)$ are defined by the equation $V(\alpha) = |E|$, so that $0 < \alpha_0 < \alpha_1 < 1$. Thus, the amplitude A of the oscillatory mode increases from $M_0\sqrt{\alpha_0}$ to $M_0\sqrt{\alpha_1}$, decreases back to $M_0\sqrt{\alpha_0}$, and then goes back and forth. At the same time, according to Eq. (4.25), the amplitude B of the spring mode decreases and increases correspondingly.

4.1.2   *Chaotic behavior of a spring pendulum*

The analysis of the sequence of order-chaos-order transitions in the spring pendulum was performed [160] by the use of the two dimensionless control parameters, R and μ. The parameter R was defined in (4.7). The second parameter μ is defined by

$$\mu = 1 + \frac{\omega_s^2}{\hat{\omega}_0^2} \tag{4.29}$$

This parameter can range from 1 to infinity.

The equations of motion (4.4) of a spring pendulum, have been analyzed [160] for different values of the control parameters R and μ. Chaotic states appear for intermediate values of R and μ, whereas regular non-chaotic solutions occur for the limiting values of these parameters.

The different types of solutions in the $R - \mu$ plane are displayed in Fig. 4.2. The shading in the central region of this plane indicates chaotic solutions. The regular solutions shown at points a, b, e, f, whereas points c and d represent the chaotic solutions. Chaos is connected with the coupling of two degrees of freedom, so that the point $\mu = 5$, which corresponds to autoparametric resonance, is a natural source of chaos. The boundary between the locked and running solutions is another region where the chaotic solutions are clustered. For a rigid pendulum, the boundary is defined by the separatrix, whereas for a spring pendulum, these boundaries are described [160] by the following curves in the $R - \mu$ plane, below which there

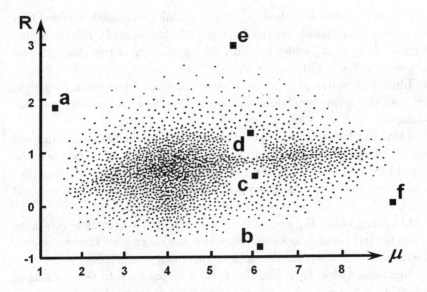

Fig. 4.2 The $(\mu - R)$-plane. The points **a**, **b**, **e**, **f** correspond to regular trajectories whereas the shadow region (points **c**, **d**) corresponds to critical trajectories.

are no running solutions,

$$R = 1 - 2\left[(2\,(\mu - 1) + 1)\right]^{-1}, \qquad \mu > 2$$

$$R = (\mu - 1)\left[2 + (\mu - 1)^{-1}\right]^{-1}, \qquad 1 < \mu < 2 \qquad (4.30)$$

At $\mu = 2$, where $\omega_s = \omega_0$, these two curves coincide showing the borderline for running solutions.

Although the original equations contain chaotic solutions, these solutions appear only in the intermediate range of the control parameters (Fig. 4.2). It is instructive to consider the several limits in which the system of equations (4.4) is integrable.

Limiting value $\mu \to 1$ can be approached when the spring is weak (small κ) or the bob is heavy (large m). One can then rewrite the equations of motion (4.12) and (4.13),

$$\frac{d^2 X}{dt^2} + \omega_s^2 X = (\mu - 1)\,gX\left[X^2 + \left(Z + l_0 + g/\omega_s^2\right)^2\right]^{-1/2} \qquad (4.31)$$

$$\frac{d^2 Z}{dt^2} + \omega_s^2 Z + g = (\mu - 1)\,gZ\left[X^2 + \left(Z + l_0 + g/\omega_s^2\right)^2\right]^{-1/2} \qquad (4.32)$$

For very small $\mu - 1$, these equations describe two weakly coupled harmonic oscillators which, in the limit $\mu = 1$, have smooth elliptic orbits. For $\mu = 1$, $\hat{\omega}_0 = \omega_s$, which explains the appearance of this characteristic frequency in Eqs. (4.31) and (4.32).

Limiting value $\mu \to \infty$ **or** $\kappa \to \infty$ marks the transition to a rigid rod, i.e., the spring pendulum becomes a simple pendulum with periodic solutions.

Limiting value R $\to (-1)$ corresponds to small oscillations (librations) near the downward position. In this case, the equations of motion take the form (4.18) and (4.19), which define two coupled oscillators. However, for small deviations from the downward position, the coupling is very small which leads to simple solutions for harmonic oscillators.

Limiting value R $\to \infty$ **or E** $\to \infty$ results in two equivalent rotations, clockwise and counterclockwise, for which the energies of the two modes, elastic and oscillatory, are the same order of magnitude [160].

Limiting value E $\to -\infty$ (holds for $g \to \infty$) leads to the splitting of equations (4.4) for $x \ll y$, which again yields an integrable system.

4.1.3 *Driven spring pendulum*

In previous sections, we considered the properties of an insolated spring pendulum. An extension of the pendulum model consists of adding an additional external periodic force of frequency Ω. The equations of motion in polar coordinates (r, ϕ) can be easily obtained from the Lagrangian,

$$\frac{d^2 r}{dt^2} + c_1 \frac{dr}{dt} + \kappa r - (1+r) \left(\frac{d\phi}{dt}\right)^2 + \frac{g}{l}(1 - \cos\phi) = 0 \qquad (4.33)$$

$$(1+r)^2 \frac{d^2\phi}{dt^2} + c_2 \frac{d\phi}{dt} + 2(1+r)\frac{dr}{dt}\frac{d\phi}{dt} + \frac{g}{l}(1+r)\sin\phi = 0 \qquad (4.34)$$

An external force can be added to Eqs. (4.33) and (4.34) in two different ways. An external force can act on the bob which leads to an additive term,

$$\frac{d^2 r}{dt^2} + c_1 \frac{dr}{dt} + \kappa r - (1+r) \left(\frac{d\phi}{dt}\right)^2 + \frac{g}{l}(1 - \cos\phi) = K \cos(\Omega t) \qquad (4.35)$$

$$(1+r)^2 \frac{d^2\phi}{dt^2} + c_2 \frac{d\phi}{dt} + 2(1+r)\frac{dr}{dt}\frac{d\phi}{dt} + \frac{g}{l}(1+r)\sin\phi = L \cos(\Omega t) \qquad (4.36)$$

Another possibility corresponds to a periodic force inducing vertical oscillations $u(t)$ of the suspension point. As explained in Section 2.3, the suspension point has acceleration d^2u/dt^2 relative to our inertial frame of reference. One introduces a non-inertial frame having this acceleration. Then, the gravity constant g in the Lagrangian (2.18) has to be replaced by $g + d^2u/dt^2$. For $u(t) = A\cos(\Omega t)$, Eqs. (4.33) and (4.34) are replaced by the following equations [159],

$$\frac{d^2r}{dt^2} + c_1\frac{dr}{dt} + \kappa r - (1+r)\left(\frac{d\phi}{dt}\right)^2 + \frac{g}{l}(1 - \cos\phi)$$

$$-\frac{A\Omega^2}{l}\cos(\Omega t)\cos\phi = 0 \tag{4.37}$$

$$(1+r)^2\frac{d^2\phi}{dt^2} + c_2\frac{d\phi}{dt} + 2(1+r)\frac{dr}{dt}\frac{d\phi}{dt} + \frac{g}{l}(1+r)\sin\phi$$

$$+\frac{A\Omega^2}{l}\cos(\Omega t)\sin\phi = 0 \tag{4.38}$$

An external periodic signal $\cos(\Omega t)$ enters Eqs. (4.35) and (4.36) additively and Eqs. (4.37) and (4.38) multiplicatively. To obtain an approximate solution of these systems of nonlinear differential equations, one uses the method of multiple scales, described in Section 2.3.7. Such an analysis of Eqs. (4.35) and (4.36) has been discussed previously (see [163], [164] and references therein), and has been recently extended [165] to the case of vertical oscillations of the suspension point, described by Eqs. (4.37) and (4.38). We will not write down the cumbersome forms of the linearized equations, restricting ourselves to the numerical solutions of Eqs. (4.35) and (4.36) for the onset of chaos in the special cases of external resonance ($\Omega = \omega_0 \equiv \sqrt{g/l}$ and $\Omega = \omega_s \equiv \sqrt{\kappa}$) and internal autoparametric resonance ($\omega_s = 2\omega_0$), where ω_0 and ω_s denote the linearized natural frequencies of the pendulum and spring modes, respectively. To describe the frequencies near the resonance conditions, we introduce detuning parameters σ_1 and σ_2,

$$2\omega_0 = \omega_s + \sigma_1; \qquad \Omega = \omega_0 + \sigma_2 \text{ and } \Omega = \omega_s + \sigma_2 \tag{4.39}$$

External spring-mode resonance ($\Omega \approx \omega_s$) [163]. For $\omega_0 = 1, c_1 = c_2 = 0.005, \sigma_1 = 0.03, \sigma_2 = 0.004, L = 0$, changing the value of the control parameter K causes the pendulum motion to change from the limit circles for $K = 0.000364$ to chaotic motion for $K = 0.00055$. This result has been obtained from the Poincare map by inspecting the phase projection at times $t = n2\pi/\Omega$ with integer n, and calculating the largest Lyapunov exponent. Chaotic motion appears through a sequence of period-doubling bifurcations (see Fig. 2.13).

External pendulum-mode resonance ($\Omega \approx \omega_0$) [163]. Similar results have been obtained in this case for the parameters $\omega_s = 0.5$, $c_1 = c_2 = 0.005$, $\sigma_1 = -0.01$, $\sigma_2 = 0.005$, $K = 0$, going from $L = 0.0007$ to $L = 0.000873$.

By calculating the Lyapunov exponents for different values of the damping parameters c_1, c_2 and the detuning parameters σ_1, σ_2 and comparing analytical results with the numerical results, it was found [166] that the second-order approximation of the multiple scales method leads to better agreement with the numerical results than the first-order approximation. The comprehensive analysis for different values of control parameters c_1, c_2 and σ_1, σ_2 has been performed [164] for the spring-mode resonance (with $\omega_s = 1$ and $K = 0.0055$) and for the pendulum-mode resonance (with $\omega_0 = 0.5$ and $L = 0.00087$). The bifurcation diagrams and Poincare maps show that there is only one equilibrium state upon changing the internal tuning parameter σ_1 while the number of equilibrium states varies between three and one upon a change of the external detuning parameter σ_2. For some deviation from the resonance condition and for some regimes of damping parameters, these equilibrium states become unstable and the solutions become quasiperiodic or chaotic.

Equations (4.35) and (4.36) have recently been generalized [167]-[169]. The nonlinear spring stiffness term $\kappa_i r^i$ ($i = 2, 3, 4$) has been included in Eq. (2.68). Numerical calculations show that the steady state solutions and the onset of chaos are monotonically decreasing functions of the damping coefficients. A generalization of Eqs. (4.35) and (4.36) implies that the bob is forced to move in a circular path of radius r with frequency ω [168]. Numerical calculations have been performed for the following parameters: $\omega_s = 0.5$, $\omega = 0.5$, $\sigma_1 = 0.01$, $\sigma_2 = 0.005$, $c_1 = c_2 = 0.05$, $l = 1$ and $K = 0$. Both the phase diagram (r as a function of Ωt) and the $r - \phi$ Poincare map show a sequence of period-doubling bifurcations leading to chaotic motion for four values of r (0.03, 0.07, 0.10 and 0.20). Analysis of Eqs. (4.35) and (4.36) shows that changing the control parameter (amplitude L of the external force) leads to a series of quasiperiodic solutions followed by the onset of chaos. These authors supplemented the nonlinear coupling between the radial and angular motion of the pendulum by a second source of nonideality powered by an additional $\kappa_3 r^3$ term in Eq. (4.35) (nonlinearity of the elastic spring of the pendulum). To compare these two types of nonlinearity required an analysis of Poincare maps and the calculation of the Floquet multipliers and the Lyapunov exponents. The former reveals the geometrical structure of the attractors whereas the latter indicates the nature of the route to chaos. Although the addition of the second nonlinearity did not change the route to chaos, the chaotic motion occurs in different frequency regimes. Moreover, the second nonlinearity could promote the stabilization of the periodic motion beyond the threshold of instability. There is a distinction between a weakly nonlinear spring (the parameter κ_3 being small) and strongly nonlinear spring (κ_3 is close to unity). In the former case, the addition of a small linearity has a stabilizing effect on the motion, extending the range for which the motion remain periodic, i.e., the onset of chaos occurs at larger amplitudes of an external field than for the case of a linear spring. However, quite surprisingly, the strong nonlinearity of the spring stabilizes the pendulum in a certain frequency range and destabilizes the pendulum for other values of frequency.

4.2 Double pendulum

In the previous section, we considered the chaotic behavior of an elastic pendulum with two degrees of freedom, the angle ϕ and the length r. Another nonlinear system which has two degrees of freedom is the double pendulum, which consists of two simple pendulums attached to each other (Fig. 4.3).

The position of a double pendulum is characterized by two angles ϕ_1 and ϕ_2 between the rods and the vertical. The length of the rods l_1 and l_2 and the bobs masses m_1 and m_2 are the control parameters. The positions of the bobs are

$$x_1 = l_1 \sin \phi_1; \quad y_1 = -l_1 \cos \phi_1; \quad x_2 = l_1 \sin \phi_1 + l_2 \sin \phi_2;$$
$$y_2 = -l_1 \cos \phi_1 - l_2 \cos \phi_2 \tag{4.40}$$

The Lagrangian of a double pendulum is

$$L = \frac{1}{2} (m_1 + m_2) l_1^2 \left(\frac{d\phi_1}{dt} \right)^2 + \frac{1}{2} m_2 l_2^2 \left(\frac{d\phi_2}{dt} \right)^2 \tag{4.41}$$
$$+ m_2 l_1 l_2 \left(\frac{d\phi_1}{dt} \right) \left(\frac{d\phi_2}{dt} \right) \cos (\phi_1 - \phi_2) + (m_1 + m_2) g l_1 \cos \phi_1 + m_2 g l_2 \cos \phi_2$$

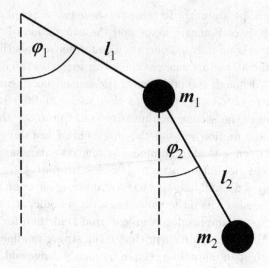

Fig. 4.3   The double pendulum. Reprinted from [170], Copyright (2006), with permission from Elsevier.

leading to the following equations of motion

$$(m_1 + m_2) \, l_1 \frac{d^2\phi_1}{dt^2} + m_2 l_2 \frac{d^2\phi_2}{dt^2} \cos\left(\phi_1 - \phi_2\right)$$

$$+ m_2 l_2 \left(\frac{d\phi_2}{dt}\right)^2 \sin\left(\phi_1 - \phi_2\right) + (m_1 + m_2) \, g \sin\phi_1 = 0 \qquad (4.42)$$

$$m_2 l_2 \frac{d^2\phi_2}{dt^2} + l_1 \frac{d^2\phi_1}{dt^2} \cos\left(\phi_1 - \phi_2\right) - l_1 \left(\frac{d\phi_1}{dt}\right)^2 \sin\left(\phi_1 - \phi_2\right) + g \sin\phi_2 = 0$$

$$\qquad (4.43)$$

For small oscillations, $\sin\phi \approx \phi$ and $\cos\phi \approx 1$, the normal modes $\phi_{1,2} = \phi_{1,2}^{(0)} \exp\left(\omega t\right)$ are

$$\omega^2 = \frac{g \left(l_1 + l_2\right)}{2 m_1 l_1 l_2} \left\{ m_1 + m_2 \pm \sqrt{(m_1 + m_2) \left[m_2 + m_1 \left(\frac{l_1 - l_2}{l_1 + l_2}\right) \right]^2} \right\}$$

$$\qquad (4.44)$$

$$\frac{\phi_2^{(0)}}{\phi_1^{(0)}} = \frac{g - \omega^2 l_2}{\omega^2 l_1} \qquad (4.45)$$

If the masses and lengths of two pendula are equal, $m_1 = m_2$ and $l_1 = l_2$,

Eqs. (4.44) and (4.45) simplify to

$$\omega = \sqrt{\frac{g}{l}}\sqrt{2 \pm \sqrt{2}}; \qquad \frac{\phi_2^{(0)}}{\phi_1^{(0)}} = \frac{-1 \mp \sqrt{2}}{2 \pm \sqrt{2}} \qquad (4.46)$$

The latter equations are the solution of the linearized equation of a mathematical pendulum with two masses moving in phase or out of phase.

The case without gravity ($g = 0$) or two uncoupled pendula ($m_2 = 0$) are also integrable. The original equations (4.42) and (4.43) are more complicated than equations (4.12) and (4.13) for the elastic pendulum, and one has to resort to a numerical solution [170]. Since this system is Hamiltonian (no damping), the energy is conserved, and is given by the initial conditions. In two limits, zero and infinite energy, the motion is regular. The difficulty is to find chaotic behavior for finite, non-zero values of the energy. The conserved energy is given from the Lagrangian (4.41),

$$E\left(\phi_1, \phi_2, \frac{d\phi_1}{dt}, \frac{d\phi_2}{dt}\right) = \left(\frac{d\phi_1}{dt}\right)\frac{\partial L}{\partial(d\phi_1/dt)} + \left(\frac{d\phi_2}{dt}\right)\frac{\partial L}{\partial(d\phi_2/dt)} - L$$

$$(4.47)$$

Numerical calculations have been performed [170] for various values of energy, $E = -9, -7, 7$, and 9, corresponding to initial conditions $(0, 0, 0, 0)$, $(0, \pi, 0, 0)$, $(\pi, 0, 0, 0)$, and $(\pi, \pi, 0, 0)$, respectively. The phase-space sections have been obtained for initial conditions in the interval $(-0.01, 0.01)$ around the initial conditions given above. The phase space is four dimensional. In order to obtain a plane Poincare section, two parameters have been fixed, $\phi_1 = 0$ and $d\phi_1/dt > 0$. The Poincare sections for different values of energy are shown in Figs. 4.4–4.6

The behavior becomes increasingly non-regular as the energy increases. Figure 4.4, corresponding to energy $E = -9$, presents a very regular image. The motion becomes more complex for $E = -0.7$ (Fig. 4.5), and finally, for $E = 7$ (Fig. 4.6) almost all trajectories are chaotic. These result have been confirmed [170] by calculating the Lyapunov exponents.

A comprehensive analysis of Eqs. (4.42) and (4.43) has been performed [171]. It was shown that an analytic solution is possible only if the three parameters, m_2/m_1, l_2/l_1 and E/m_1gl_1 are equal to zero or infinity. The Melnikov method was applied to the different limiting cases.

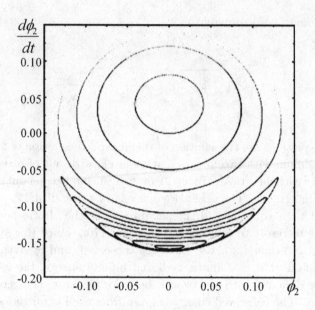

Fig. 4.4   Poincare section for $E = -9 + 0.01$, regular trajectories. Reprinted from [170], Copyright (2006), with permission from Elsevier.

4.3   Spherical pendulum.

Another version of a pendulum with two degrees of freedom (angles ϕ and θ) is shown in Fig. 4.7. The bob of mass m, suspended on the rod of length l, is able to swing in any direction performing two-dimensional motion on the surface of a sphere of radius l, described by the angles ϕ and θ. Placing the origin at the suspension point and using spherical coordinates with the z-axis directed downward, the bob coordinates are $x = l \sin\theta \cos\phi$, $y = l \sin\theta \sin\phi$, $z = -l\cos\theta$. The Lagrangian L is

$$L = \frac{1}{2}ml^2\left[\left(\frac{d\theta}{dt}\right)^2 - \left(1 - \cos^2\theta\right)\left(\frac{d\phi}{dt}\right)^2\right] + mgl\cos\theta, \qquad (4.48)$$

which leads to the following equations of motion,

$$ml^2\left(\frac{d^2\theta}{dt^2}\right) - ml^2\sin\theta\cos\theta\left(\frac{d\phi}{dt}\right)^2 + mgl\sin\theta = 0 \qquad (4.49)$$

$$ml^2\sin\theta\left(\frac{d^2\phi}{dt^2}\right) + 2ml^2\sin\theta\cos\phi\left(\frac{d\theta}{dt}\right)\left(\frac{d\phi}{dt}\right) = 0 \qquad (4.50)$$

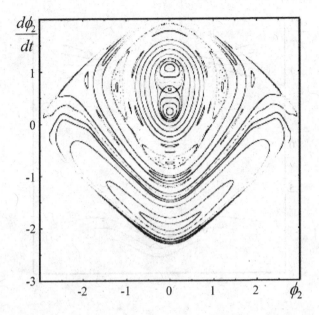

Fig. 4.5 Poincare section for $E = -7 - 0.01$, regular trajectories begin to deform. Reprinted from [170], Copyright (2006), with permission from Elsevier.

The latter equation can be rewritten

$$\frac{d}{dt}\left[ml^2 \sin^2\theta\left(\frac{d\phi}{dt}\right)\right] \equiv \frac{dJ}{dt} = 0 \qquad (4.51)$$

which yields the conservation of J, the angular moment in the ϕ-direction.

Another conserved quantity is the energy

$$E = \frac{1}{2}\left(\frac{d\theta}{dt}\right)^2 + V_{eff}(\theta); \quad V_{eff}(\theta) = -\frac{g}{l}\cos\theta + \frac{J^2}{2m^2l^4\sin^2\theta} \qquad (4.52)$$

In contrast to the spring and the double pendulum, the spherical pendulum has two conserved quantities. The motion of the bob $\theta(t)$ is restricted by the two solutions of equation $E = V_{eff}(\theta)$, so that $\theta_1 < \theta < \theta_2$. The values of the turning points θ_1 and θ_2, where $d\theta/dt = 0$, depend on the parameters E and J, which are given by the initial conditions. A simple result is obtained [172] for initial conditions $\theta_0 = \pi/2$, $\phi_0 = 0$, $(d\theta/dt)_0 = 0$, $(d\phi/dt)_0 = \sqrt{2g/l}$), namely, the turning points located at height $0 < |z(t)| < l\psi$, where ψ is called the golden number $\psi = (\sqrt{5}+1)/2$.

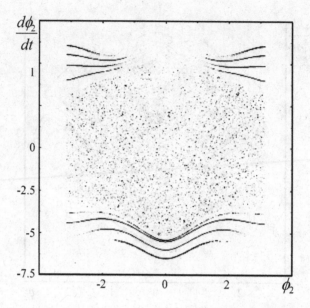

Fig. 4.6   Poincare section for $E = 7 - 0.01$, regular trajectories decay with the appearance of global, completely chaotic trajectories. Reprinted from [170], Copyright (2006), with permission from Elsevier.

In general, one finds the solution $\theta(t)$ of the equations of motion numerically [68]. Integrating Eq. (4.52) yields

$$t - t_0 = \frac{1}{\sqrt{2}} \int \frac{d\theta}{\sqrt{E - V_{eff}(\theta)}} \qquad (4.53)$$

From the solution of this equation, one finds $\phi(t)$ using Eq. (4.51),

$$\phi = \int \frac{J dt}{\sin^2 \theta \sqrt{E - V_{eff}(\theta)}} \qquad (4.54)$$

Thus far, we considered the free motion of a spherical pendulum. If the suspension point oscillates periodically in the horizontal x-direction with the (linearized) pendulum frequency $\sqrt{l/g}$ and the amplitude is smaller than the length of the rod, the bob traces an elliptic path having the driving frequency in the plane of the driving force. In the presence of the driving

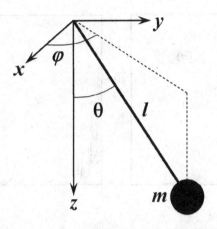

Fig. 4.7 Spherical pendulum.

force and damping, Eqs. (4.49) and (4.50) have following form [173]

$$\left(\frac{d^2\theta}{dt^2}\right) - \left[\omega_0^2 - \left(\frac{d\phi}{dt}\right)^2 \cos\theta\right] \sin\theta + \frac{\alpha}{m}\left(\frac{d\theta}{dt}\right) + \frac{1}{l}\left(\frac{d^2x}{dt^2}\right) \cos\phi\cos\theta = 0$$

$$(4.55)$$

$$\sin\theta\left(\frac{d^2\phi}{dt^2}\right) + 2\cos\theta\left(\frac{d\theta}{dt}\right)\left(\frac{d\phi}{dt}\right) + \frac{\beta}{m}\left(\frac{d\phi}{dt}\right) - \frac{1}{l}\left(\frac{d^2x}{dt^2}\right)\sin\phi = 0 \quad (4.56)$$

The approximate analytical solution of Eqs. (4.55) and (4.56) has been obtained [174] under the assumptions that the driving frequency is close to ω_0 and that the damping coefficients are equal and small, $\alpha/m = \beta/m << 1$. A slight change of the driving frequency results in pendulum motion in the plane perpendicular to the direction of the driving force with the random jumps between these two planes, thereby changing the regular motion into chaotic motion. An approximate analytic analysis, as well as numerical calculations and their experimental versification, have been carried out for the undamped pendulum at driving frequencies $0.985\sqrt{g/l}$, $0.990\sqrt{g/l}$, and $0.998\sqrt{g/l}$ [175] and for frequencies in the interval $(0.97 - 1.035)\sqrt{g/l}$ [176], as well as for the damping coefficient $\alpha/m = \beta/m = 0.01$ [177] and 0.003 [173]. All these analyses show the appearance of chaos.

As an example, we display [175] in Fig. 4.8 the power spectrum for the pendulum being driven at its natural frequency ω_0 and at the nearby

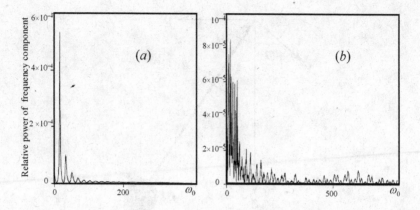

Fig. 4.8   Power spectra for pendulum with (linearized) frequency, ω_0, and driven frequency, ω, for a) $\omega = \omega_0$ and b) $\omega = 0.998\omega_0$. Reprinted with permission from [175],
Copyright (2001), Institute of Physics.

frequency 0.998 ω_0. In the former case, one obtains one very large peak
and a few smaller peaks which decay. The power spectrum at 0.998 ω_0
looks very different, showing a typical chaotic type of motion in the form
of a series of large peaks in the low-frequency range and a broad spread of
smaller peaks across the rest of the frequency range.

Chapter 5

Conclusions

The pendulum is the simplest nonlinear system which contains the characteristics of different phenomena in Nature in the areas of physics, chemistry, biology, engineering, economics and sociology. Based on experiment, the modern scientist creates a new theory for making predictions, which eventually have to be tested experimentally. In Brownian motion, the random collisions of the test particle by the surrounding molecules is responsible for the random motion of the Brownian particle. There are also other sources of randomness in Nature, such as thermal fluctuations which always exist or the random changes in the surrounded medium. Therefore, there is a need to add random forces (additive in the former case and multiplicative in the latter case) to the differential equations describing such phenomena.

Another problem appears in the macroscopic description of many-body systems. Although, as Laplace said, knowing the initial conditions of 10^{23} particles permits one to make complete predictions of their dynamic behavior in the future, this approach is clearly not practical. Therefore, one abandons the description of the individual particles in favor of the average characteristics of the system as a whole.

A new understanding of the properties of nonlinear differential equations is connected with the phenomenon of deterministic chaos, which shows the deep relationship between deterministic and random phenomena, and, like relativity and quantum mechanics, lies far beyond the scope of science forming an important part of our worldview. Due to the exponential sensitivity to initial conditions, which are only known to restricted accuracy, even a simple deterministic system, such as the periodically driven pendulum, is unpredictable. Such random-like behavior already occurs in a driven, nonlinear system with two degrees of freedom!

This book contains the comprehensive description of the random (chaotic) behavior of a pendulum subject to different combination of constant, periodic and random forces. The different impact of these forces manifests itself, for example, in suppressing chaos by the common action of multiplicative and additive forces, whereas each force by itself induces chaos. Likewise, chaos can be suppressed by the addition of random forces to the equation which predicts chaos in the absence of a random force. Still another possibility for suppressing chaos is by adding a second weak periodic force. The existence of chaos may be harmful for some applications and, therefore, the suppression of chaos is of great interest.

The model of a pendulum is widely used in modern science. In addition to the different applications described in Section 1.5, there are many other applications, including chemical reactions [178; 179], biophysics (neural activity [180], intracellular transport [119], oscillations in the visual cortex [181], penetration of biological channels by ions [182]), superionic conductors [183], plasma physics [184], surface diffusion [185], electrophoresis [186], rotation of molecules in solids [187], dielectric relaxation [188], and polymer dynamics [189], engineering (ship dynamics [190]), gravitational gradient pendulum [191], and matter-antimatter asymmetry in the universe [192]. However, these and other applications of the pendulum model rarely contain the analysis of chaotic behavior. This subject still attracts great interest, and many new applications will surely be found in the future.

Bibliography

[1] M. Gitterman, Eur. J. Phys. **23**, 119 (2002).
[2] M. Gitterman, *The Noisy Pendulum* [World Scientific, 2008].
[3] E. I. Butikov, Eur. J. Phys. **20**, 429 (1999).
[4] L. Reichl, *The Transition to Chaos* [Springer, 1992].
[5] E. T. Whittaker and G. N. Watson, *A Course of Modern Analysis* [Cambridge University Press, 1927].
[6] I. Lira, Eur. J. Phys. **28**, 289 (2007).
[7] R. R. Parwani, Eur. J. Phys. **25**, 37 (2004).
[8] Y. Ging-Xin and D. Pei, Eur. J. Phys. **30**, L79 (2009).
[9] M.I. Molina, Phys. Teach. **35**, 489 (1997).
[10] R. B. Kidd and S. L. Fogg, Phys. Teach. **40**, 81 (2002).
[11] F. M. S. Lima and P. Arun, Am. J. Phys. **74**, 892, (2006).
[12] A. Cromer, Am. J. Phys. **63**, 112 (1995).
[13] C. G. Carvalhaev and P. Suppers, Am. J. Phys. **76**, 1150 (2008).
[14] G. L. Baker and J. P. Gollub, *Chaotic Dynamics, an Introduction* [Cambridge University Press, 1990].
[15] A. H. Nayfeh and D. T. Mook, *Nonlinear Oscillations* [Wiley, 1979].
[16] D. Permann and I. Hamilton, Am. J. Phys. **60**, 442 (1992).
[17] P. Coullet, J. M. Gilli, M. Monticelli, and N. Vandenberghe, Am. J. Phys. **73**, 1122 (2005).
[18] N. F. Pedersen and O. H. Sorrensen, Am. J. Phys. **45**, 994 (1977).
[19] H. Risken, *The Fokker-Planck Equation* [Springer, 1996].
[20] B. Ya. Shapiro, M. Gitterman, I. Dayan, and G. H. Weiss, Phys. Rev. B **46**, 8349 (1992).
[21] R. Adler, Proc. IREE **34**, 351 (1946).
[22] A. Pikovsky, M. Rosenblum, and J. Kurtis, *Synchronization - A Universal Concept in Nonlinear Science* [Springer, 2002].
[23] M. Gitterman, Phys. Rev. A **35**, 41 (1987); M. Gitterman and D. Pfeffer, Synthetic Metals **18**, 759 (1987); J. Magn. Magn. Materials **68**, 243 (1987); M. Gitterman, Thermodyn. Acta **169**, 47 (1989).
[24] C. F. Bak and N. F. Pedersen, Appl. Phys. Lett. **22**, 149 (1973).
[25] M. Cirillo and N. F. Pedersen, Phys. Lett. A **90**, 150 (1982).

[26] J. Kallunki, M. Dube, and T. Ala-Nissila, Surf. Sci. **460**, 39, (2000).

[27] R. Guanfes, J. L. Vega, and S. Miret-Artes, Phys. Rev. B **64**, 245415, (2001).

[28] O. M. Brown and Y. S. Kivshar, Phys. Rep. **306**, 1 (1998).

[29] Y. V. Kartashov, L. Torner and A. Vysloukh, Opt. Lett. **29**, 1102 (2004).

[30] X. Xu, *Nonlinear Dynamics of the Parametric Pendulum for Wave Energy Extraction* [University of Aberdeen, 2006].

[31] B. H. Suits, Eur. J. Phys. **27**, L7 (2006).

[32] J. Falzano, A. T. Roesh, and H. Troger, in *Bifurcation and Chaos: Analysis, Algorithms and Applications*, vol. 97, p. 117 [Birkhaeuser-Verlag, Basel, 1991].

[33] A. Wolf and T. Bessoir, Physica D **50**, 239 (1991).

[34] R. L. Kautz and B. M. Huggard, Am. J. Phys. **62**, 59 (1994).

[35] R. Carretero-Gonzalez, N. H. Nunez-Yepez, and A. L. Salas-Brito, Eur. J. Phys. **15**, 139 (1994).

[36] R. Cuerno, A. F. Ranada, and J. Ruiz-Lorenzo, Am. J. Phys. **60**, 73 (1992).

[37] J. P. Eckmann, Rev. Mod. Phys. **53**, 643 (1981).

[38] Y. Pomeau and P. Manneville, Comm. Math. Phys. **74**, 189 (1980); P. Manneville and Y. Pomeau, Phys. Lett. A **75**, 296 (1980).

[39] M. Octavio, Physica A **163**, 248 (1990).

[40] D. D'Humueres, M. R. Beasley, B. A. Huberman, and A. Libchaber, Phys. Rev. A **26**, 3483 (1982).

[41] B. A. Huberman, J. R. Crutchfield, and N. H. Packard, Appl. Phys. Lett. **37** 750 (1980).

[42] J. A. Blackburn, Z. J. Yang, S. Vik, H. J. T. Smith, and M. A. H. Nerenberg, Physica D **26**, 385 (1987).

[43] N. F. Pedersen and A. Davidson, Appl. Phys. Lett. **39**, 830 (1981).

[44] R. L. Kautz, J. Appl. Phys. **86**, 5794 (1999).

[45] W. J. Yeh and Y. H. Kao, Appl. Phys. Lett **42**, 299 (1983).

[46] E. G. Gwin and R. M. Westervelt, Phys. Rev. Lett. **54**, 1613 (1985); Phys. Rev. A **33**, 4143 (1986).

[47] A. H. MacDonald and M. Plischke, Phys. Rev. B **27**, 201 (1983).

[48] Z.. Abbadi and E. Simiu, Nanotechnology, **13**, 153 (2002).

[49] Y. Braiman and I. Goldhirsch, Phys. Rev. Lett. **66**, 2545 (1991).

[50] D.-R. He, W. J. Yeh, and Y. H. Kao, Phys. Rev. B **30,** 172 (1984).

[51] N. Takimoto and M. Tange, Progr. Theor. Phys. **90**, 817 (1992).

[52] W. C. Kerr, M. B. Williams, A. R. Bishop, K. Fesser, P. S. Lomdahl, and S. E. Trullinger, Z. Phys. B **59**, 103 (1985).

[53] F. Palmero and F. R. Romero, Phys. Lett. A **160**, 553 (1991).

[54] G. Ambika, J. Phys.: Condens, Matter, **4**, 4829 (1992).

[55] K. I. Thomas and G. Ambika, Pramana **59**, 445 (2002).

[56] R. Fitzpatrick, *Chaotic Pendulum*, http://farside.ph.utexas.edu/teaching/329/lectures/node54.html.

[57] R. Harish and K. P. N. Marthy, Proc. Second Nation. Conf. on Nonlin. Systems and Dynamics [Aligarh, Muslim Univ. 2006].

[58] J. A. Blackborn and N. Gronbech-Jensen, Phys. Rev. E **53**, 3068 (1996).

[59] M. Feigenbaum, J. Stat. Phys. **19**, 25 (1978).

[60] T. Can and W. Moore, *Fractals and Chaos in the Driven Pendulum*, http://hep.unchicago.edu/~rosner/p316/projs/can_moore.pdf.

[61] M. Feigenbaum, Los Alamos Science **1**, 41 (1978).

[62] J Awrejcewicz and M. M. Holicke, *Smooth and Non-Smooth High Dimensional Chaos and the Melnikov-Type Methods* [World Scientific, 2007].

[63] G. Grebogi, E. Ott, and J. A. Yorke, Physica D **24**, 243 (1987).

[64] T. Schwalger, A. Dzhanoev, and A. Loskutov, Chaos **16**, 023109, (2006).

[65] V. K. Melnikov, Trans. Moscow Math. Soc. **12**, 1 (1963).

[66] M. Bartuccelli, P. L. Christiansen, N. F. Pedersen, and M. P. Soerensen, Phys. Rev. B **33**, 4686 (1986).

[67] J. B. McLaughlin, J. Stat. Phys. **24**, 375, (1981).

[68] L. D. Landau and E. M. Lifshitz, *Mechanics* [Butterworth-Heinemann, 1976].

[69] R. W. Leven and B. P. Koch, Phys. Lett. A **86**, 71 (1981).

[70] B. P. Koch, R. W. Leven, B. Pompe, and C. Wilke, Phys. Lett. A **96**, 219 (1983).

[71] S. R. Bishop and M. J. Clifford, J. Sound Vibr. **189**, 142 (1996).

[72] E.-A. Kim, K.-C. Lee, M. Y. Choi, and S. Kim, J. Korean Phys. Soc. **44**, 518 (2004).

[73] B. P. Koch and R. W. Leven, Physica D **16**, 1 (1985).

[74] M. A. F. Sanjuan, Phys. Rev. E **58**, 4377 (1998).

[75] M. A. F. Sanjuan, Chaos, Solitons and Fractals, **9**, 995 (1998).

[76] A. Stephenson, Mem. Proc. Manchester Lit. Phil. Soc. **52**, 1 (1908).

[77] E. Butikov, Am. J. Phys. **69**, 755 (2001).

[78] T. Leiber and H. Risken, Phys. Lett. A **129**, 214 (1988}.

[79] M. V. Bartuccelli, G. Gentile, and K. V. Georgiou, Proc. Roy. Soc. A **457**, 3007 (2001).

[80] M. I. Clifford and S. R. Bishop, J. Sound Vibr. **172**, 572, (1994).

[81] S. Lenci, E Pavlovskaia, G. Rega, and M Wiercigroch, J. Sound Vibr. **310**, 243, (2008).

[82] R. van Dooren, Chaos, Solitons and Fractals, **7**, 77 (1996).

[83] P. J. Bryant and J. W. Miles, J. Austral. Math. Soc. Ser. B, **34**, 153 (1992).

[84] J. M. Schmitt and P. V. Bayly, Nonlin. Dynam. **15**, 1 (1998).

[85] A. H. Nayfeh and B. Balachandran, *Applied Nonlinear Dynamics* [Wiley, 1995].

[86] J. L. Trueba, J. . Baltanas, and M. A,.F. Sanjuan, Chaos, Solitons and Fractals **15**, 911 (2003).

[87] B. Wu and J. A. Blackburn, Phys. Rev. A **45**, 7030 (1992).

[88] H. J. T. Smith and J. A. Blackburn, Phys. Rev. A **40**, 4708 (1989).

[89] F.-G. Xie and W.-M. Zheng, Phys. Rev. E **49**, 1888, (1994).

[90] J. A. Blackburn, H. J. T.Smith and D. E. Edmundson, Phys. Rev. A **45**, 593 (1992).

[91] H. Seifert, Phys. Lett. A **98**, 213 (1983).

[92] R. L. Kautz and R. Monako, J. Appl. Phys. **57**, 875 (1985).

[93] G. Cicogna, Phys. Lett. A **121**, 403 (1987).

[94] C. Vanneste, C. C. Chi, and D. C. Cronemeyer, Phys. Rev. B **32**, 4796 (1985).

[95] X. Yao, J. Z. Wu, and C. S. Ting, Phys. Rev. B **42**, 244 (1990).

[96] M. Bikdash, B. Balachandran, and A. Nayfeh, Nonlin. Dynam. **6**, 101 (1994).

[97] S.-Y. Kim and K. Lee, Phys. Rev. E **53**, 1579 (1996); S.-Y. Kim, S.-H. Shin, J. Yi, and C.-W. Lang, Phys. Rev. E **56**, 6613 (1997); S.-Y. Kim and B. Hu, Phys. Rev. E **58**, 3028 (1998).

[98] A. Sofroniou and S. R. Bishop, Chaos, Solitons and Fractals, **28**, 673 (2006).

[99] P. Zhou and H. Cao, Chaos, Solitons and Fractals, **38**, 590 (2008).

[100] Z. Jing and J. Yang, Int. J. Bifurc. Chaos **16**, 2887 (2006).

[101] D. W. Jordan and P. Smith, *Nonlinear Ordinary Differential Equations* [Oxford, Clarendon Press 1987].

[102] I. W. Stewart and T. R. Faulkner, Phys. Rev. E **62**, 4856 (2000).

[103] B. Bruhn and B. R. Koch, Z. Naturforsch. A **43**, 930 (1988).

[104] V. E. Shapiro and V. M. Loginov, Physica A **91**, 563, (1978).

[105] N. G. van Kampen, *Stochastic Processes in Physics and Chemistry* [North-Holland, 1992].

[106] M. Gitterman, R. I. Shrager, and G. H. Weiss, Phys. Lett A **142**, 84 (1989).

[107] S. Chandrasekar, Rev. Mod. Phys. **15**, 1 (1943).

[108] Yu. M. Ivanchenko and L. A. Zil'berman, Soviet Phys.-JETP **55**, 2395 (1968); V. Ambegaokar and B. I. Halperin, Phys. Rev. Lett. **22**, 1364 (1969).

[109] P. Fulde, L. Petronero, W. R. Schneider, and S. Strassler, Phys. Rev. Lett. **35**, 1776 (1975).

[110] M. Borromeo, G. Constantini, and F. Marchesoni, Phys. Rev. Lett. **82**, 2820 (1999).

[111] K. Lindenberg, J. M. Sancho, A. M. Lacaste, and I. M. Sokolov, Phys. Rev. Lett. **98**, 020602 (2007).

[112] R. Harrish, S. Rajasekar, and K. P. N. Murthy, Phys. Rev. E **65**, 046214 (2002).

[113] B. Chen and J. Dong, Phys. Rev. B **44**, 10206 (1991).

[114] N. F. Pedersen, Physica Scripta, **T13**, 129 (1986).

[115] R. F. Miracky and J. Clarke, Appl. Phys. Lett. **43**, 508 (1983).

[116] M. N. Popescu, Y. Braiman, F. Family, and H. G. E. Hentschel, Phys. Rev. E **58**, R4057 (1998).

[117] T. Kapitaniak, Phys. Lett. A **116**, 251, (1986).

[118] M. F. Weher and W. G. Wolfer, Phys. Rev. A **27**, 2663 (1983).

[119] P. Reinmann, Phys. Rep. **361**, 57 (2002).

[120] P. Jung, J. S. Kissner, and P. Hanggi, Phys. Rev. Lett. **76**, 3436 (1996).

[121] J. L. Mateos, Phys. Rev. Lett. **84**, 258 (2000).

[122] F. R. Alatriste and J. L. Mateos, Physica A **384**, 233 (2007).

[123] L. Machura, M. Kostur, P. Talkner, J. Luczka, and P. Hanggi, Phys. Rev. Lett. **98**, 040601 (2007).

[124] D. Speer, R. Eichhorn, and P. Reimann, Europhys. Lett. **79**, 1005 (2007); Phys. Rev. E **76**, 051110 (2007).

[125] K. Mallick and P. Marcq, J. Phys. A **37**, 4769 (2004).

[126] K. Mallick and P. Marcq, Phys. Rev. E **66**, 041113 (2002).

[127] J. A. Blackburn, Proc. Roy. Soc. London A **462**, 1043 (2000).

[128] J. A. Blackburn, N. Gronbeck-Jensen, and H. J. T. Smith, Phys. Rev. Lett. **74**, 908 (1995).

[129] G. Litak, M. Borowiec, and M. Wiercigroch, Dynamic Systems, **23**, 259 (2008).

[130] U. Lepik and H. Hein, J. Sound Vibr. **288**, 275 (2005).

[131] H. Hein and U. Lepik, J. Sound Vibr. **301**, 1040 (2007).

[132] L. R. Nie and D. C. Mei, Eur. Phys. J. B **58**, 475 (2007).

[133] R. L. Stratonovich, Radiotech. Electronic (in Russian) **3**, 397 (1958); R. L. Stratonovich, *Topics in the Theory of Random Noise* [Gordon and Breach, 1967].

[134] P. Reimann, C. Van den Broek, H. J. Linke, P. Hanggi, M. Rubi, and A. Perez-Madrid, Phys. Rev. E **65**, 031104 (2002).

[135] S. Lifson and J. J. Jackson, J. Chem. Phys. **36**, 2410 (1962).

[136] L. S. Gradstein and I. M. Ryzhik, *Tables of Integrals, Series and Products* [Academic, 1994].

[137] M. Gitterman, I. B. Khalfin, and B. Ya. Shapiro, Phys. Lett. A **184**, 339 (1994).

[138] E. Heinsalu, R. Tammelo, and T. Ord, arXiv: cond-mat/0208532v2.

[139] M. M. Millonas and D. B. Chialvo, Phys. Rev. E **53**, 2239 (1996).

[140] V. Berdichevsky and M. Gitterman, Phys. Rev. E **56**, 6340 (1997).

[141] Van den Broek, J. Stat. Phys. **31**, 467 (1983); J. Luczka, R. Bartussek, and P. Hanggi, Europhys. Lett. **31**, 431 (1995).

[142] C. R. Doering and J. C. Gadoua, Phys. Rev. Lett. **69**, 2138 (1992).

[143] V. Berdichevsky and M. Gitterman, Physica A **249**, 88 (1998).

[144] G. W. Gardiner, *Handbook of Stochastic Methods* [Springer, 1997].

[145] I. I. Fedchenia and N. A. Usova, Z. Phys. B **50**, 263 (1983).

[146] V. Berdichevsky and M. Gitterman, Phys. Rev. E **65**, 011104 (2001).

[147] H. Horsthemke and R. Lefever, *Noise-Induced Phase Transitions* [Springer, 1984].

[148] A. J. R. Madureira, P. Hanggi, V. Buonomano, and W. A. Rodrigues, Jr., Phys. Rev. E **51**, 3849 (1995).

[149] M. Marchi, F. Marchesoni, L. Gammaitoni, E. Manichella-Saetta, and S. Santucci, Phys. Rev. E **54**, 3479 (1996).

[150] S. H. Park, S. Kim, and C. S. Ryu, Phys. Lett. A **225**, 245 (1997).

[151] S. L. Ginzburg and M. A. Pustovoit, Phys. Rev. Lett. **80**, 4840 (1998); JETP **89**, 801 (1999); O. V. Gerashchenko, S. L. Ginzburg and M. A. Pustovoit, Eur. J. Phys. B **15**, 335 (2000); **19**, 101 (2001).

[152] C. Zhou and C.-H. Lai, Phys. Rev. E **59**, R6243 (1999); **60**, 3928 (1999).

[153] G. H. Weiss and M. Gitterman, J. Stat. Phys. **70**, 93 (1993).

[154] M. R. Young and S. Songh, Phys. Rev. A **38**, 238 (1988).

[155] M. Gitterman, J. Phys. A **32**, L293 (1999).

[156] C.-J. Wang, S.-B. Chen, and D.-C. Mei, Phys. Lett. **352**, 119 (2006).

[157] J.-H. Li and Z.-Q. Huang, Phys. Rev. E **58**, 139 (1998).

[158] Y.-H. Luo and Z.-W. Xie, Phys. Lett. A **373**, 3217 (2009).

[159] A. Arinstein and M. Gitterman, Eur. J. Phys. **29**, 385 (2008).

[160] J. P. van der Weele and E. de Kleine, Physica A **228**, 245 (1996).

[161] N. Minorsky, *Nonlinear Oscillations* [Van Nostrand, 1962].

[162] H. M. Lai, Am. J. Phys. **52**, 219, (1984).

[163] W. K. Lee and H. D. Park, Nonlinear Dynamics **14**, 211 (1997).

[164] A. Alasty and R. Shabani, Nonlinear Analysis: Real World Applications **7**, 81, (2006).

[165] M. Gitterman, Physica A **389**, 3101 (2010).

[166] W. K. Lee and H. D. Park, Int. J. of Nonlin. Dynamics, **34**, 749(1999).

[167] M. Eissa, S. A. El-Serafi, M. El-Sheikh, and M. Sayed, Appl. Math and Comput. **145**, 421, (2003).

[168] T. S. Amer and M. A. Bek, Nonlinear. Analisys: Real World Applications **10**, 3196, (2009).

[169] K. Zaki, S. Noah, K. R. Rajagopal, and A. R. Srinivasa, Nonlin. Dynamics **27**, 1 (2002).

[170] T. Stachowiak and T. Okada, Chaos, Solitons & Fractals. **29**, 417 (2006).

[171] A. V. Ivanov, Regular and Chaotic Dynamics, **4**, 104 1999 ; J. Phys. A **34**, 11011 (2001).

[172] Essen and N. Apazidis, Eur. J. Phys. **30**, 427 (2009).

[173] D. D. Kana and D. J. Fox, Chaos, **5**, 298 (1995).

[174] J. Miles, Physica D **11**, 309 (1984).

[175] S. J. Martin and P. J. Ford, Phys, Education **36**, 108 (2001).

[176] D. J. Tritton, Eur. J. Phys. **7**, 162 (1986).

[177] P. J. Bryant, Physica D **64**, 324 (1993).

[178] Y. Kuramoto, *Chemical Oscillations, Waves and Turbulence* [Springer, 1984].

[179] V. Petrov, Q. Ouyand, and H. L. Swinney, Nature **388**, 655 (1997).

[180] C. Kurrer and K. Shulten, Phys. Rev. E **51**, 6213 (1995).

[181] H. Sompolinsky, D. Golomb, and D. Kleinfeld, Phys. Rev. A **43**, 6990 (1991).

[182] A. J. Viterby, *Principles of Coherent Communications* [McGraw-Hill, 1966].

[183] W. Dietrich, P. Fulde, and I. Peschel, Adv. Phys. **29**, 527 (1980).

[184] E. M. Lifshitz and L. P. Pitaevskii, *Physical Kinetics* [Pergamon, 1981].

[185] P. Talkner, E. Hershkovitz, E. Pollak, and P. Hanggi, Surf. Sci. **437**, 198 (1999).

[186] C. L. Asbury and G. van den Engh, Biophys. J. **74**, 1024 (1998).

[187] Y. Georgievskii and A. S. Burstein, J. Chem. Phys. **100**, 7319 (1994).

[188] J. R. McConnel, *Rotational Brownian Motion and Dielectric Theory* [Academic, 1980].

[189] G. I. Nixon and G. W. Slater, Phys. Rev. E **53**, 4969 (1996).

[190] K. J. Spyrou, Phil. Trans. Roy. Soc. Ser. A **358**, 1733 (2000).

[191] S. W. Ziegler and M. P. Cartmell, J. Spacecraft Rockets **38**, 904 (2001).

[192] M. E. Shaposhnikov, Contemp. Phys. **39**, 177 (1998).

Glossary

Attractor. Irrespective of initial conditions, all trajectories reach the attractor as $t \to \infty$.

Basin boundary. Boundary between two (or more) sets of initial conditions leading to different attractors.

Basin of attractors. The locus of all points in phase space which lead to this attractor.

Bifurcation. Change in the behavior of a dynamic system upon change of control parameter(s).

Bifurcation diagram. Describes the dynamic behavior of a system for different values of the control parameters.

Center. Equilibrium position surrounded by closed phase trajectories (ellipses).

Crisis. Sudden change of attractors induced by a change of control parameters.

Elliptic point. All nearby trajectories have the form of ellipses.

Feigenbaum transition to chaos. The cascade of period-doubling bifurcations leading to chaos.

Fixed point. A point which is mapped onto itself after iteration of the phase map.

Fractals. An object that displays self-similarity (repetition of the structure) on all scales.

Harmonic balance method. Method of approximate solution of a differential equation by the use of periodic functions.

Homoclinic trajectories. Trajectories that begin and end at a saddle point surrounding the center.

Hopf bifurcation. Transition from an equilibrium state to a point attractor or to a limit cycle induced by a change of control parameters.

Hyperbolic point. All nearby trajectories have the form of a hyperbola.

Limit cycle. The periodic solution of a differential equation plotted in phase space.

Lyapunov exponent. Number defining the exponential approach or divergence of two neighboring trajectories with slightly different initial conditions. Positive Lyapunov exponent corresponds to chaotic motion.

Periodic attractor. Attractor which corresponds to motion that is periodic in time.

Poincare section. An undriven pendulum showing periodic sampling of state parameters ϕ and $d\phi/dt$.

Subcritical bifurcation. A point in the phase space where the steady-state solution becomes unstable.

Saddle point. Point for which trajectories are stable for one direction, but unstable for other directions.

Separatrix. Curve separating elliptic (closed) and hyperbolic (open) trajectories.

Strange attractor. Attractor which has fractal structure.

Supercritical bifurcation. A point in phase space where the solution changes its form but remains stable.

Trajectory. Solution of differential equation with given initial conditions plotted in phase space.

Transient chaos. Motion which looks chaotic for some finite time interval, but afterwards returns to non-chaotic motion.

A comprehensive glossary of concepts and terms in the study of nonlinear mechanics and chaos can be found in "The Illustrated Dictionary of Nonlinear Dynamics and Chaos," by T. Kapitaniak and S R. Bishop [Wiley, 1999].

Index